Power Systems

Electrical power has been the technological foundation of industrial societies for many years. Although the systems designed to provide and apply electrical energy have reached a high degree of maturity, unforeseen problems are constantly encountered, necessitating the design of more efficient and reliable systems based on novel technologies. The book series Power Systems is aimed at providing detailed, accurate and sound technical information about these new developments in electrical power engineering. It includes topics on power generation, storage and transmission as well as electrical machines. The monographs and advanced textbooks in this series address researchers, lecturers, industrial engineers and senior students in electrical engineering.

More information about this series at http://www.springer.com/series/4622

Gaber Magdy • Gaber Shabib
Adel A. Elbaset • Yasunori Mitani

Renewable Power Systems Dynamic Security

Springer

Gaber Magdy
Electrical Engineering Department
Faculty of Energy Engineering
Aswan University
Aswan, Egypt

Gaber Shabib
Electrical Engineering Department
Faculty of Energy Engineering
Aswan University
Aswan, Egypt

Adel A. Elbaset
Electrical Engineering Department
Faculty of Engineering
Minia University
El-Minia, Egypt

Yasunori Mitani
Electrical Engineering Department
Kyushu Institute of Technology
Tobata-ku, Kitakyushu-shi
Fukuoka, Japan

ISSN 1612-1287 ISSN 1860-4676 (electronic)
Power Systems
ISBN 978-3-030-33454-3 ISBN 978-3-030-33455-0 (eBook)
https://doi.org/10.1007/978-3-030-33455-0

This Springer imprint is published by the registered company Springer Nature Switzerland AG
The registered company address is: Gewerbestrasse 11, 6330 Cham, Switzerland

Dedicated to our families and students

Abstract

Concerning the environmental issues, energy crisis, and economic growth, the utilization of renewable energy sources (RESs) towards high penetration in our community is seriously inevitable. For instance, the Ministry of Electricity and Renewable Energy of Egypt plans to increase the electric energy from RESs to cover 20% of the electricity demand by 2020 and 42% of the electricity demand by 2030. Thus, the popularity of RESs throughout the world will increase rapidly due to the three aforementioned issues. However, most of the RESs are connected to the power system through the power electronic interfaces called inverters/converters. The utilization of inverters/converters will significantly reduce the inertia of the power system (community). With the trend to increase the RESs in the power system, the overall inertia of the power system will significantly reduce and create many stability problems in system frequency and voltage, leading to the weakening of the power system. This problem will widely affect people's daily life and community by power interruption and, in the worst case, may lead to power failure/blackout. The low system inertia issue is also one of the major restrictions to integrate RESs, which are clean, cheap, and sustainable to all communities. Therefore, the renewable power systems (RPSs) have become more susceptible to the system insecure than traditional power systems because of the following; reducing of the overall inertia of the power system that result from replacing the conventional generators, e.g., synchronous generators (SGs) with RESs, and decoupling of the RESs from the AC grid using power converters. Thus, maintaining the dynamic security of RPSs is the key challenge for integrating more RESs. Hence, this book proposes new frequency control techniques based on superconducting magnetic energy storage (SMES) system, virtual inertia control, and virtual synchronous generator (VSG) for frequency stability enhancement of RPS considering the high penetration level of RESs. However, a suitable inertia control technique can be applied together with energy storage systems (ESSs) to emulate additional inertia power to the community or power system, improving system inertia and eliminating the stability issues. Moreover, the proposed frequency control strategies are coordinated with digital over/underfrequency protection for the enhancement of frequency stability and preservation of the dynamic security of RPSs because of the high integration level of RESs. The effectiveness of the proposed coordination schemes is tested and verified through small and large scales of RPSs, i.e., Microgrid (μG) and Egyptian Power System (EPS). The simulation results proved that RPSs with the proposed coordinated schemes will provide better stability and performance for today's power system and for those of the future, which are expected to integrate more and more RESs; thus, the proposed coordination schemes will ensure avoidance of power system instability and system collapse.

Acknowledgments

First and foremost, all our thankfulness is to Allah who helped and guided us to carry out this work. We would like to take this opportunity to extend our heartfelt appreciation to the following persons who have contributed directly or indirectly towards the completion of the book.

The authors would like to thank Prof. Dr. Hassan Bevrani (University of Kurdistan, Sanandaj, Iran), Dr. Yaser Qudaih (American University, Madaba, Jordan), and Dr. Thongchart Kerdphol (Kyushu Institute of Technology, Kitakyushu, Japan) for their active role and continuous support. Last but not least, the authors offer their deepest personal gratitude to their family for all their patience and help during the preparation of this book.

Contents

Nomenclatures

P^n_{gd}	gbest of particle i at iteration n
P^n_{id}	pbest of particle i at iteration n
v^n_{id}	The velocity of particle i at iteration n
$x_i^{\max}$	The upper limit
$x_i^{\min}$	The lower limit
x^n_{id}	The current position of particle i at iteration n
Δf	The system frequency deviation (Hz)
ΔP_{C}	Regulating the microgrid frequency (Hz)
ΔP_{c1}	Regulating the system frequency of non-reheat plant (Hz)
ΔP_{c2}	Regulating the system frequency of reheat plant (Hz)
ΔP_{c3}	Regulating the system frequency of hydro plant (Hz)
ΔP_{d}	Change in demand power (pu)
$\Delta P_{\mathrm{Hydro}}$	Power deviation of hydraulic power plant (pu)
$\Delta P_{\mathrm{Inertia}}$	Inertia power change (pu)
ΔP_{L}	Load variation (MW pu)
ΔP_{m}	Power deviation of thermal power plant (pu)
$\Delta P_{\mathrm{Non\text{-}Reh.}}$	Power deviation of non-reheat power plant (pu)
ΔP_{PV}	Solar farm power change (pu)
$\Delta P_{\mathrm{Reh.}}$	Power deviation of reheat power plant (pu)
ΔP_{SMES}	The active power deviation of SMES unit
ΔP_{WT}	Wind farm power change (pu)
$1 - g/G$	The cognitive factor
$2g/G$	The social factor
A	The state matrix
A_{T}	The rotor swept area (m^2)
B	Corresponds to the control input signal
$\mathrm{best_g}$	The global best solution
$\mathrm{best_p}$	The best light source position
C	Corresponds to the output measurement
c_1, c_2	Acceleration constant

D	Zero vector with the same size of input control signal
$D_{\mu G}$	Microgrid damping coefficient (pu MW/Hz)
D_{EPS}	System damping coefficient of the EPS (pu MW/Hz)
D_i	System damping coefficient of the virtual rotor (pu MW/Hz)
D_s	Load damping coefficient (pu MW/Hz)
E	Corresponds to the disturbance inputs
f	System frequency (Hz)
F	Zero vector with the same size of the input disturbance vector
f_{Max}	Maximum frequency limit (Hz)
f_{Min}	Minimum frequency limit (Hz)
GRC	Generation rate constraint, % (pu)
$H_{\mu G}$	Microgrid system inertia (pu MW s)
H_{EPS}	Equivalent inertia constant of the EPS (pu s)
H_i	Equivalent inertia constant of the virtual rotor (pu s)
K	Integrator set time
K_1	Virtual primary proportional gain
K_2	Virtual secondary integrator gain
K_d	Derivative control variable gain
K_i	Integral control variable gain
K_p	Proportional control variable gain
K_{SMES}	SMES variable gain
K_{VI}	Virtual inertia control gain
m	The fraction of turbine power (intermediate pressure section)
n	Number of iterations
n_o	The size of onlooker moths
P_{n1}	Nominal rated power output for the non-reheat plant (MW pu)
P_{n2}	Nominal rated power output for reheat plant (MW pu)
P_{n3}	Nominal rated power output for the hydro plant (MW pu)
P_{Wind}	The output power of the wind turbine (W)
R	Droop constant (Hz/pu MW)
R_1	Governor speed regulation non-reheat plant (Hz/pu MW)
R_2	Governor speed regulation reheat plant (Hz/pu MW)
R_3	Governor speed regulation hydro plant (Hz/pu MW)
rand ()	A random number between 0 and 1
r_T	The rotor radius
T	Sampling time interval (s)
T_1	The valve time constant of the non-reheat plant (s)
T_2	The steam valve time constant of reheat plant (s)
T_3	Water valve time constant hydro plant (s)
Td	Communication time delay (s)
T_d	The dashpot time constant of hydro plant speed governor (s)
T_g	The time constant of the governor (s)
T_h	The time constant of reheat thermal plant (s)
T_{IN}	Inverter time constant (s)

T_{PV}	Solar system time constant (s)
T_s	Maximum settling time (s)
T_{SMES}	SMES time constant (s)
T_t	The time constant of the turbine (s)
T_{VI}	Time constant-based virtual inertia (s)
T_w	Water starting time in hydro intake (s)
T_{WT}	Wind turbine time constant (s)
U	The control input signal
V_L	The minimum limit of valve gate (pu MW)
V_U	The maximum limit of valve gate (pu MW)
V_W	The rated wind speed (m/s)
w	Inertia weight factor
W	The input disturbance vector
X	The state vector
Y	The control output signal
β	The pitch angle
δ	Rotor angle (degree)
ε_1	Random samples were drawn from Gaussian stochastic
ε_2, ε_3	Random numbers within the interval [0,1]
ρ	Air density (kg/m^3)
λ	The tip speed ratio (TSR)
xxx_o	Synchronous speed (rpm)
AGC	Automatic generation control
DGs	Distributed generators
EPS	Egyptian Power System
ESS	Energy storage system
EVs	Electric vehicles
FDE	Frequency detection element
FLC	Fuzzy logic control
FMU	Frequency measurement unit
GDB	Governor dead-band
GRC	Generation rate constraint
HWPP	High wind power penetration
ISE	Integral of squared error
LFC	Load frequency control
MOS	Maximum overshoot
MPC	Model predictive control
MSA	Moth swarm algorithm
MUS	Maximum undershoot
OUFR	Over/Underfrequency relay
PID	Proportional–integral–derivative controller
PSO	Particle swarm optimization
PSS	Power system stabilizer
PV	Photovoltaic

RESs	Renewable energy sources
RoCoF	Rate of change of frequency
RPS	Renewable power system
SG	Synchronous generator
SMES	Superconducting magnetic energy storage
VSG	Virtual synchronous generator

List of Figures

List of Tables

Chapter 1
Introduction and Literature Review

1.1 An Overview and Motivations

In recent years, most world countries face the great challenge of modernizing an aging grid infrastructure. Furthermore, the programs of RPSs, which are considered environmentally friendly, are on many government agendas. Therefore, the use of new and renewable sources of energy such as wind, solar, and geothermal has turned out to be inescapable. Hence, the face of today's power system is changing due to many reasons, such as environmental concerns, energy system security, fossil fuel problems, and economical and operation cost issues. Many countries around the world have decided to increase the penetration level of RESs in their energy system. However, the intermittent nature of the RESs causes many control problems such as increasing the power imbalance in the short-term operation of power systems, and frequency/voltage instability problem, which may be limiting their high penetration [1]. Moreover, the RESs exchange electrical power with the RPSs through power electronic devices (i.e., inverters and converters), which are static devices. Therefore, most of the RESs lack a rotating mass, which is the main source of inertia, and thus the associated inertia constant is roughly zero [2]. Hence, by increasing the penetration level of RESs into RPSs, the influence of low system inertia and damping effect on the dynamic system performance and stability increases. Furthermore, this low system inertia issue could affect the power system stability and resiliency in the situation of uncertainties, and thus threaten their dynamic security. Consequently, the inverter-based RESs will cause high frequency/voltage fluctuations compared to the conventional generation units. As a result, RPSs become more susceptible to the disturbances than traditional power systems, and thus are facing some of the disturbances that threaten their dynamic security such as large frequency/voltage fluctuations, sudden load shedding, forced islanding incidents, and short-circuit faults with long clearing times [3].

With the continuous development in electrical loads particularly industrial plants and human activities, there has been an increased number of new transmission lines,

G. Magdy et al., *Renewable Power Systems Dynamic Security*, Power Systems,
https://doi.org/10.1007/978-3-030-33455-0_1

power plants, and interconnection between different power systems. This leads to the appearance of the frequency and power oscillation problems as well as tie-line power deviation in the interconnected power system, which may result in disconnection actions, loss of several lines, zone islanding, equipment damaging, transmission line overload, and interference with system protection schemes [4]. Nowadays, this problem increases after growing the RESs, which have several impacts on the performance of the electrical power systems such as reduction of the overall system inertia that results to increase the frequency and voltage fluctuations [5]. The aforementioned reasons directly affect the power system operation, stability, and security. One of the most important indexes of power systems is the voltage frequency. Recently, frequency control in power systems has gained considerable attention due to its importance. Therefore, load frequency control (LFC) is considered as one of the most important control strategies in the power system, which maintains the system frequency and the power variations at their standard values. Whereas system frequency depends on active power, system voltage greatly depends on the reactive power. Therefore, the control of power systems can be classified into two fundamental issues: (a) control of the active power along with the frequency (i.e., LFC) and (b) control of the reactive power along with the voltage regulation [6]. So, the main objectives of the LFC are [7]:

- Regulating the system frequency and tracking the load demands
- Maintaining frequency and power interchanges with neighboring control areas at the specified values
- Ensuring zero steady-state error for frequency deviations
- Controlling the change in the tie-line power between control areas

This literature review highlights the LFC models in different power systems, which are divided into two main groups (i.e., LFC models in conventional power systems, and LFC models in modern and future power systems). Furthermore, the control approaches for the LFC in power systems are surveyed.

1.2 Types of Power System Models

The traditional power system that has been in use since centuries from the generation and transmission level to the distribution was mainly dominated by thermal, gas, hydro, and nuclear power generation, which is an integral component of the traditional power system. However, the use of renewable energy increased greatly just after the first big oil crisis in the late 1970s. At that time, economic issues were the most important factors; hence the interest in such processes decreased when oil prices fell. The current resurgence of interest in the use of renewable energy is increasing due to global warming. It affects humans in several aspects such as economics, public health, and environment. The global warming is caused by greenhouse gases, which come from burning fossil fuels such as oil or coal [7]. Harvesting energy on a large scale is undoubtedly one of the main challenges

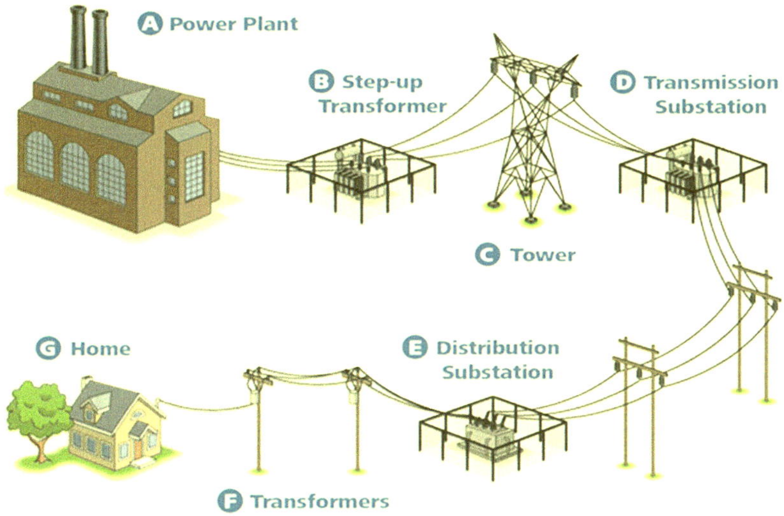

Fig. 1.1 A simplified diagram of a conventional power system [9]

of our time. Future energy sustainability depends heavily on how the renewable energy problem is addressed in the next few decades. The energy demand in the world is steadily increasing and new types of energy sources must be found in order to cover future demands since the conventional sources are about to be emptied. Therefore, the nonconventional energy sources play a vital role, in the form of distributed generator- (DG)/RES-based power system. Due to its importance in LFC studies, frequency response modeling of different power system structures is firstly reviewed. LFC models in different power systems can be divided into two main groups: conventional and modern LFC models [8].

1.2.1 Conventional Power Systems

The conventional power system refers to the electric power system in which the electricity is generated from fossil energy sources as shown in Fig. 1.1, where thermal units, hydropower plant, and nuclear generating units are the well-known power plants for such systems [8]. Thus, the conventional structures of LFC models are surveyed based on the configurations of power systems. Consequently, frequency response models of signal-area, two-area, three-area, and four-area power systems are comprehensively reviewed. In the literature, several frequency response models are suggested for LFC in which a comprehensive survey regarding power system models for LFC is given in the following subsections.

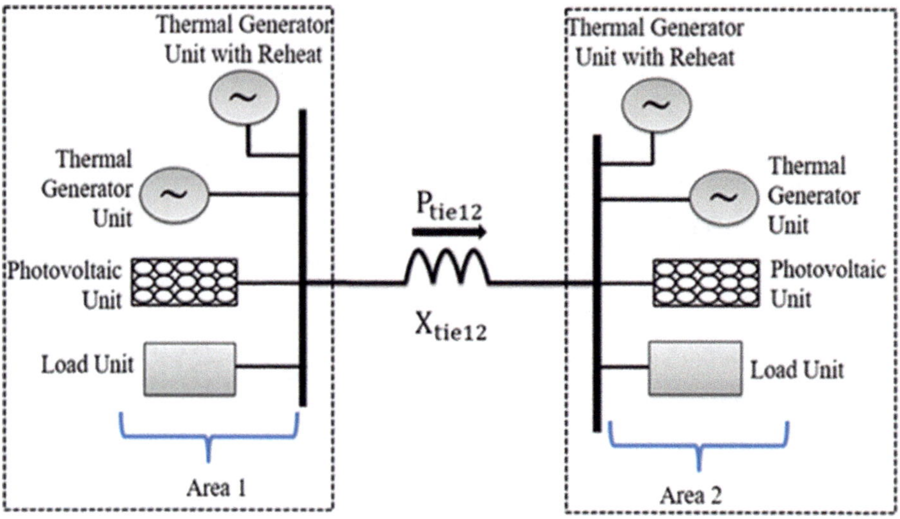

Fig. 1.2 Two-area interconnected power system [27]

1.2.1.1 Single-Area Power Systems

Many types of system models have been presented considering different generation types, such as thermal, gas, and hydropower plants. Several models of a single-area power system incorporating LFC control schemes are investigated in [10–16]. The LFC problem of single-area thermal power systems is presented in [10–12]. In [11], the LFC scheme of single-area thermal power system considering generation rate constraint (GRC) is presented. The LFC scheme of single-area thermal power system considering time delay is marked out in [12]. The LFC problem of single-area hydropower systems is presented in [13, 14]. An automatic generation control (AGC) system for a single-area hydropower system considering some nonlinearities is suggested in [14]. The single-area power systems with multisource dynamic generators, thermal, gas, and hydropower plants, incorporating LFC control schemes are presented in [15, 16].

1.2.1.2 Tow-Area Power Systems

An overview of LFC system in two-area power systems is presented in [17–26]. Figure 1.2 shows a two-area power system, where each area supplies to its own area and the power flow between the areas is allowed by the tie-line. LFC models for two-area power systems considering the GRC nonlinearities and governor dead-band (GDB) are suggested in [17, 18]. The LFC scheme of two-area power system considering the communication delays is proposed in [19]. LFC models of two-area power system taking into account the effect of system uncertainties are marked out in [19, 20]. The frequency control strategy for two-area power systems tied together via

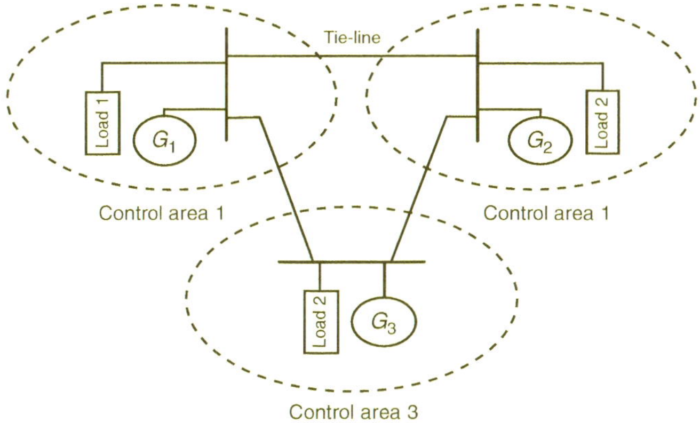

Fig. 1.3 Three-control-area power system [34]

HVDC/DC transmission links is presented in [21]. The LFC models of multisource two-area power system considering nonlinearities are presented in [22]. The frequency response model for a two-area power system with SMES system is proposed in [23]. Two-area power system considering the contribution of SMES and batteries has been presented for frequency stability analysis in [24]. The interconnected two-area power system with GRC and boiler dynamics including SMES units for LFC is considered in [25]. A frequency control model for RPS considering the RES uncertainties is proposed in [26].

1.2.1.3 Three-Area Power Systems

The frequency control strategies for a three-area power system are investigated in [28–33]. A typical model of the three-area interconnected power system and its individual subsystems is shown in Fig. 1.3. A frequency control model for a three-area interconnected power system is given in [28]. LFC model for three-area power systems considering the GRC and GDB nonlinearities is presented in [29, 30]. The effects of the communication delays on frequency response model in three-area interconnected power systems are addressed in [31]. Furthermore, the effects of system uncertainties on LFC in a three-area power system are highlighted in [32, 33]. The LFC model for multisource power systems in which thermal, gas, and hydropower plants are considered is proposed in [33].

1.2.1.4 Four-Area Power Systems

LFC challenges in four-area interconnected power systems are presented in [35–38]. The LFC model for a four-area interconnected power system with four identical thermal units is presented in [35]. LFC model for four-area power systems considering the GRC and GDB nonlinearities is presented in [36]. The system uncertainties of power system parameters are taken into account by using a fuzzy control for an LFC model in [37]. The frequency control model for a four-area interconnected power system with different turbine units in which non-reheat power plants are considered in the first and second areas while the third and fourth areas have hydropower plants is given in [38].

1.2.2 Modern and Future Power Systems

RESs are growing rapidly and highly penetrated in modern power systems. Therefore, frequency response models for power systems considering several RESs are well surveyed. Thus, the modern and future LFC models can be divided into three main groups: LFC for power systems with RESs, LFC models in microgrids (μGs), and LFC models in smart grids.

1.2.2.1 Power Systems with RESs

During the past few years, there is a growing interest in integrating RESs into the electrical grids as a future solution for the reduction of greenhouse gas emissions generated by conventional power plants, such as carbon dioxide and nitrogen oxide, which have bad outcome in the environment [3]. The LFC problems for power systems considering several RESs are addressed in [39–42]. The frequency stability support from doubly fed induction generator driven by wind turbine is proposed in [40]. Furthermore, the impacts of wind power generation on system frequency are highlighted in [41]. LFC models suitable for power systems with high penetration of RESs are presented in [39, 42].

1.2.2.2 Microgrids

The concept of μG provides an attractive solution to face the great challenges of integration of RESs into electrical grids. The μGs are small-scale power systems that interconnect DGs, ESSs, and multiple customers as shown in Fig. 1.4. Consequently, several SGs, which are considered the source of the dynamics in conventional power systems, are being replaced by the concept of DGs, which can be classified into two types: dispatchable sources such as gas turbine, microturbine, and small hydroelectric power plant, and non-dispatchable sources such as solar, wind, and wave energy [43]. The LFC problems for μGs, which consisted of photovoltaic (PV), wind turbine generation (WTG), and microturbine, are investigated in [44–46].

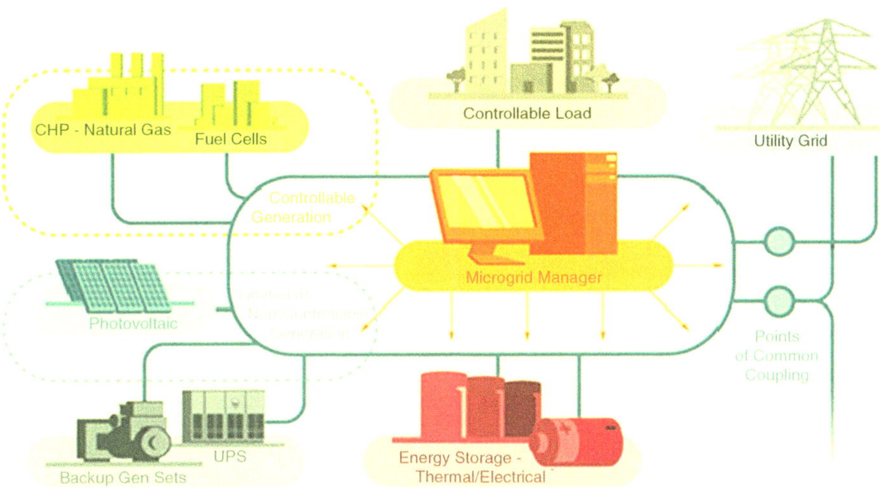

Fig. 1.4 Schematic diagram of a typical microgrid [50]

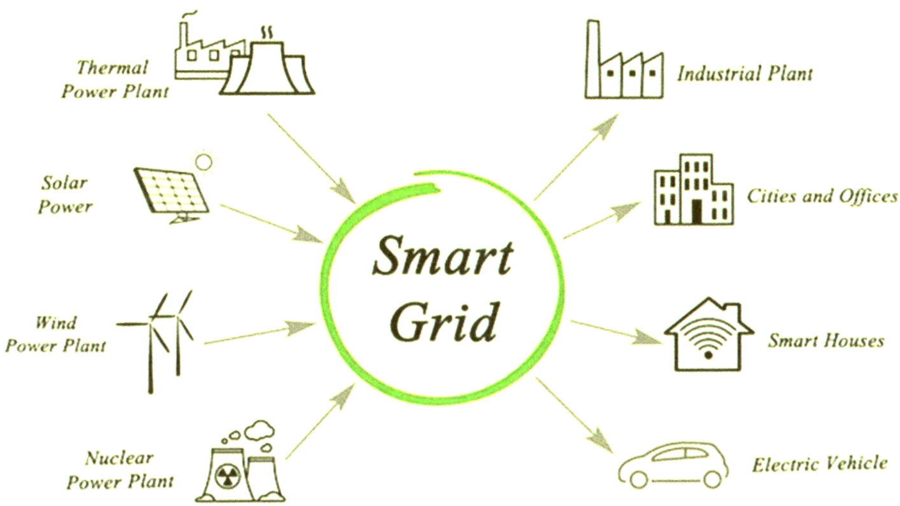

Fig. 1.5 Schematic diagram of the smart grid [51]

1.2.2.3 Smart Grids

Smart grid is built on many of the technologies already used by electric utilities but adds communication and control capabilities that will optimize the operation of the entire electrical grid as shown in Fig. 1.5. Smart grid is also positioned to take advantage of new technologies, such as plug-in hybrid electric vehicles (EVs), various forms of distributed generation, solar energy, smart metering, lighting management systems, distribution automation, and many more. Recently, the frequency control issues for smart grids have gained considerable attentions from researchers due to its great advantages. Therefore, several new LFC schemes for smart grid system have been investigated in [47–49]. The LFC model structure for smart power grids is presented in [47]. LFC models for smart grids considering the contribution of EVs are proposed in [48]. The frequency control models for future smart grids considering different ESSs have been developed in [48, 49].

1.3 Control Approaches for LFC in Power Systems

Various control techniques have been implemented for frequency stability of power systems. Control techniques can be classified into different categories: classical control techniques, optimal control approaches, adaptive control schemes, and robust control approaches. A completed survey on the proposed control approaches for LFC in power systems is presented in the following subsections.

1.3.1 Classical Control Techniques

Conventional control techniques are based on classical controllers usually applied to the governor to minimize the area control error in power systems to enhance the frequency response. Several classical control methods have been presented for LFC of power systems [6, 22, 52–58]. The performance of different classical control approaches, proportional–integral (PI), integral-double-derivative, integral, and proportional–integral–derivative (PID) controller, has been tested in [52]. The LFC strategy based on the proportional–integral plus double-derivative controller has been proposed for the multi-area power system in [53]. Various optimization algorithms were applied in the LFC loops of the interconnected power systems to obtain the optimal design of the PID parameters. Conventional methods, such as interior-point algorithm [54], aggregation method by Aoki [55], and tracking approach [56], were implemented for LFC issue. Recently, meta-heuristic optimization algorithms were used for obtaining the optimal parameters of the PID controller in the LFC system, such as particle swarm optimization (PSO) [6], chaotic optimization algorithm [57], and genetic algorithm [58]. Although these algorithms can successfully present good results where their objective functions have the capability of weighting the output signal in different objectives simultaneously, they suffer from long computational time and need several control parameters for their implementation.

1.3.2 Optimal Control Approaches

The optimal control approaches have provided solutions for multivariable control systems. These kinds of approaches consider the state variable representation of the model and an objective function to be minimized. Hence, optimal control approaches are appropriate if all state variables are available to design the feedback control signals. If the system state vector is observable with some measurements of the control area, this requirement can be met. These requirements make optimal control method complicated and undesirable for the large-scale power system. A state variable model and regulator problem of optimal control theory are used to develop a new feedback control law for the two-area interconnected power system in [59]. In [60], optimal linear regulator theory is presented to design a linear regulator for the LFC in power systems. In [61], a more realistic model of the LFC system is developed and studied under different load conditions considering the voltage-regulator excitation system and optimal responses.

1.3.3 Adaptive Control Schemes

The controller performance in the power system may not be optimal if the operating point of the system changes. Hence, to maintain the system performance near the optimal value, it would be preferable to track the operating point of the system and accordingly update its parameter to achieve a better control performance. Generally, adaptive control schemes can be classified into self-tuning control (STC) approaches and model reference control schemes. Therefore, the self-tuning regulator strategy implemented for adaptive LFC seems to be a viable solution. Various adaptive control and STC schemes have been executed for frequency stability of power systems in [4, 11, 62–64]. An adaptive fuzzy gain scheduling scheme for PID controller and optimal LFC is presented in [4]. In [11], an adaptive controller based on a PI adaptation has been proposed for LFC of power system considering system uncertainties. A multi-area adaptive LFC developed for a comprehensive AGC simulator is proposed in [62]. The STC scheme for the LFC problem of an interconnected power system is proposed in [63]. A self-tuning steam turbine control scheme designed to improve the quality of the LFC of the power system is presented in [64].

1.3.4 Robust Control Approaches

The frequency control strategies address some challenges of uncertainties and changes in system parameters (i.e., system uncertainties) and characteristics. Moreover, the uncertainties of RESs and random loads occurred in modern power systems. Therefore, the robust control design approach is applied to provide better performance (i.e., robustness) to deal with changes in the system parameters and other uncertainties. Furthermore, the robust control techniques consider physical constraints, disturbances, and uncertainties. Thus, it provides effective control synthesis methods for dynamic systems. Several research works and studies on robust control applications for LFC of power systems have been conducted in [29, 32, 33, 44, 65–67]. A decentralized robust control scheme is proposed in [33] based on active disturbance rejection control technique. A robust virtual controller-based H_infinite technique is proposed for frequency stability analysis of the islanded μG considering high penetration level of RESs in [44]. In [65], a robust controller based on the Riccati equation is presented for LFC of the power system. A robust controller-based H_infinite technique is proposed for the LFC of the hybrid generation system in [66]. H_infinite and μ-synthesis robust control techniques are applied to improve the secondary frequency control loop (i.e., LFC) in [67].

1.4 Book Objectives and Contribution

Based on the literature survey, there are some research gaps that still need to be filled in the topic of LFC of RPSs for addressing the great challenge of high penetration level of several RESs. Furthermore, the RPSs have become more susceptible to the system insecurity than conventional power systems due to reducing of the total inertia and damping properties that result from replacing the conventional generators with RESs. Therefore, maintaining the dynamic security of RPSs is the key challenge for integrating more RESs. Hence, preserving the dynamic security of RPSs is one of the important challenges, which is addressed in this book. There has not been any research done yet on RPS dynamic security as the power system protection issue has not been considered besides the frequency control issues. Therefore, the main contribution of this book includes the following aspects:

1. A new coordination of the secondary frequency control (i.e., LFC) and the SMES technology using a new optimal PID controller-based Moth Swarm Algorithm (MSA) is proposed to enhance the frequency stability of the EPS concerning high wind power penetration (HWPP) under the impact of different load profiles, random load variation, wind power fluctuations, and system uncertainties.
2. A digital coordination scheme of LFC and over/underfrequency relay (OUFR) protection is proposed for supporting the frequency stability and protecting of the islanded μG against high-frequency deviations, which increased recently due to the high penetration of RESs, random load variations, and system uncertainty.
3. A new concept of a μG frequency control incorporation in a virtual inertia control-based optimal PI controller besides the primary frequency control (i.e., governor action) and secondary frequency control (i.e., LFC) is proposed to enhance the frequency stability of the islanded μG considering high penetration level of the RESs. Moreover, the proposed virtual inertia control based on the

optimal PI controller is coordinated with the digital OUFR for enhancing the frequency stability and maintaining the μG dynamic security due to high-level RES penetration.

4. A new frequency control strategy based on VSG, which emulates the characteristics of a real SG, is proposed in RPSs to compensate the reduction in system inertia that results from adding more RESs (e.g., non-inertia sources), thus stabilizing the system frequency during high penetration of RESs. Moreover, the proposed virtual inertia control system based on VSG is coordinated with digital frequency protection for improvement of the frequency stability and preservation of the dynamic security of RPSs because of the high share of the RESs. In addition, the effectiveness of the proposed coordination scheme is tested and verified through small-scale RPSs (e.g., μG) as well as large-scale RPSs (e.g., EPS).
5. The uncertainties of renewable power generators and random load are taken into consideration in the aforementioned suggested controller design procedure. Thus, the proposed coordination schemes will guarantee an evasion of system instability and collapse.
6. A digital model of an optimal PID controller-based PSO algorithm for decentralized LFC of the EPS is proposed to replace the traditional analog model for the sake of facing the complexity of modern power systems, reduce the cost of the implementation, and increase the reliability of the control system. Moreover, the proposed digital model of decentralized LFC gives a superior robustness and frequency stabilization effect. The robustness of the proposed digital model is investigated against different load patterns, different loading, system parameter variations, an additional wind energy uncertainty, and time delays.

1.5 Book Organization and Outline

This book is separated into seven chapters and appendix. It is organized as follows:

Chapter 1 provides a comprehensive literature survey on the topic of LFC problems in power systems. In this survey, the used LFC models for various configurations of power system models are firstly investigated and classified for both conventional and future power systems. Furthermore, the proposed control techniques that concern LFC issues are addressed and categorized into different control groups. Finally, the book objectives are described, and the contributions are explained.

Chapter 2 presents a coordination scheme of LFC and SMES system using a new optimal PID controller-based MSA to enhance the frequency stability of the EPS. The configuration and state equations of the studied EPS are explained including the modeling of the wind power generation system (WPGS), power system loads, and LFC based on SMES system. A novel intelligent searching method, namely MSA, has been presented to find the optimal design parameters of the PID controller. The background of the MSA is briefly reviewed. The model of the targeted power system (i.e., the EPS) including HWPP with inherent nonlinearities and SMES system is built using MATLAB/Simulink model. The results of the proposed coordination are validated and compared with both the optimal LFC with/without the effect of conventional SMES, which is without modifying the input control signal. Then, the result evaluation of the proposed coordination is demonstrated.

Chapter 3 proposes a new digital coordination scheme of frequency stability and protection in an islanded μG. The structure of the studied μG system with the state-space equations is presented. The discrete time model of LFC-based PID controller which is discretized using mapping technique is provided. Furthermore, the modeling, as well as the principal operation of the digital OUFR, is also provided. The simulation results of the proposed digital coordination scheme by using MATLAB/Simulink model are shown and discussed. Afterwards, the result evaluation of the proposed digital coordination is provided.

Chapter 4 proposes a new frequency control strategy based on the concept of virtual inertia control to enhance the frequency stability of the islanded μG considering high penetration level of the RESs. Moreover, the proposed virtual inertia controller is coordinated with the digital OUFR for enhancing the frequency stability and maintaining the μG dynamic security due to high-level RES penetration. A brief review of the concept of virtual inertia control is provided. The state-space equations of the studied μG with the proposed virtual inertia controller are also provided. The control strategy of the virtual inertia control technique based on the optimal PI controller, which is optimally designed using the PSO algorithm, is described. Finally, the simulation results of the studied μG with the proposed coordination scheme of the virtual inertia control based on the optimal PI controller and digital OUFR are carried out using MATLAB/Simulink® software. Later, the simulation results are approved that the proposed coordination scheme can effectively regulate the μG frequency and guarantee robust performance to preserve the dynamic security of μG with high penetration of RESs for different contingencies.

Chapter 5 proposes a new frequency control strategy based on VSG, which emulates the characteristics of a real SG in RPSs to compensate the reduction in system inertia that results from adding more RESs (e.g., non-inertia sources), thus stabilizing the system frequency during high penetration of RESs. A brief review of the concept of VSG is provided. Also, the control methodology for the virtual inertia control system based on VSG is presented. The configuration of the studied μG with the implementation of PSO algorithm for VSG design as well as the simulation results and discussions are provided. Likewise, the configuration of the real hybrid power system in Egypt with the implementation of PSO algorithm for VSG design as well as the simulation results and discussions are introduced. Finally, the simulation results proved that RPSs with the proposed coordinated scheme will provide better stability and performance for today's power system, and for those of the future, which are expected to integrate more and more renewable energy; thus, the proposed coordination scheme will ensure an avoidance of power system instability and system collapse.

Chapter 6 proposes a digital model of an optimal PID controller-based PSO algorithm for decentralized LFC of the EPS to replace the traditional analog model, as the first step for upgrading the EPS towards the smart grid. The configuration and state equations of the EPS considering wind energy are explained. The continuous-time and discrete-time models for the studied power system are presented. The simulation results of the proposed digital decentralized LFC scheme by using MATLAB/Simulink model are shown and discussed. Moreover, the admissible ranges of time delays and sampling time intervals are tested and evaluated under the decentralized LFC and remain stable and the whole system stability is observed. Afterwards, the result evaluation of the proposed digital coordination is provided.

Chapter 7 presents the general conclusions and the suggested future work of the book. Also, the whole results from the proposed coordination schemes are concluded.

References give a list of used references.

Appendix (A) provides details of wind turbines that are installed in Zafarana wind farm, Egypt.

Publications from this book give a list of published international journals and conferences that are published from this book.

Chapter 2
A New Frequency Control Strategy in Real Power Systems Considering Wind Energy

2.1 Introduction

The energy demand in the world is steadily increasing and new types of energy sources must be found in order to cover future demands since the conventional sources are about to be emptied. One of the most important RESs is wind energy, which has a lower installation cost compared to the PV system; thus, it will represent a significantly larger portion of installed electrical power from renewable energy [7]. Today, wind power generation constitutes the best application of renewable energy in the aspects of investment and research. Wind energy is developing rapidly, more and more wind farms are being connected to electrical power systems, and this has some challenges that must be addressed. Therefore, this chapter presents a real hybrid power system (i.e., the EPS), which contains both conventional power plants and wind energy for facing the frequency stability issue. On the other side, the RESs can bring significant impacts to the inertia of the system when increasing their penetration level. Hence, the total system inertia will be reduced along with increasing the penetration level of RESs, which leads to increase in the system frequency deviation. Therefore, the frequency control may be difficult in case of any mismatch between electric power generation and load demand particularly with penetration growing of RESs (e.g., wind and solar energy), which are integrated into the power system.

Recently, various control techniques were implemented to the LFC of different power systems. Yousef et al. [68] used an adaptive fuzzy logic approach-based LFC for the multi-area interconnected power system. However, the approach in [68] is dependent on the designer experience to reach the required performance. Ali et al. [69] discussed the robustness of the coefficient diagram method controller including ESS (i.e., EVs) in his control strategy in a small power system. In addition, Garasi et al. [70] studied the same control strategy in [69] for the modern power system. However, the structure of this control technique in [69, 70] is complicated, as it required more steps to get its parameters. Mohamed et al. [71] used model predictive control (MPC)-based LFC for both single- and multi-area interconnected power

G. Magdy et al., *Renewable Power Systems Dynamic Security*, Power Systems,
https://doi.org/10.1007/978-3-030-33455-0_2

systems. The predictive strategy has the advantages of fast response, simple structure, and easy handle system constraints and nonlinearities. However, it takes more time for the online calculations at each sampling time. On the other hand, most industrial applications are performed based on PID controller due to it its many merits, such as economic cost, simplicity for parameter tuning, robustness, and a successful practical controller, which can provide excellent control performance regardless of the perturbations and variations in the system parameters [72]. However, the PID controller suffers from a complicated process of parameter tuning based on trial-and-error method. In such a case, the robustness of the system is not guaranteed against further perturbations in the system parameters. Therefore, finding the optimal parameters of the PID controller is one of the important challenges, which is addressed by the power system designers.

To face this issue, various optimization algorithms were applied in the LFC loops of the interconnected power systems to obtain the optimal design of the PID parameters. Conventional methods, such as interior-point algorithm (IPA) [54], aggregation method by Aoki [55], and tracking approach [56], were implemented for LFC issue. However, these methods [54–56] suffer from stagnation, trap in local minima during the optimization process, and need numerous iterations to guarantee that the solution is converged (i.e., computationally intensive). Recently, meta-heuristic optimization algorithms were used for obtaining the optimal parameters of the PID controller in the LFC system, such as PSO algorithm [6], chaotic optimization algorithm [57], and genetic algorithm [58]. Although these algorithms can successfully present good results where their objective functions have the capability of weighting the output signal in different objectives simultaneously, they suffer from long computational time and need several control parameters to implement. Moreover, the previous control strategies of LFC were implemented on the interconnected power systems, which are considered to be a linear and simple structure. Further, these power systems depend mainly on conventional power plants. Therefore, RESs should be integrated into the interconnected power system to achieve a realistic study as described in this chapter.

A few studies have dealt with the effect of incorporation RESs, particularly wind farms in the power system controlled by different LFC strategies. Mohamed et al. [34] studied the predictive control strategy based on the LFC for interconnected power system concerning wind turbines. Bhatt et al. [73] studied the effect of the dynamic contribution of the doubly fed induction generator-based wind turbine for the two-area interconnected hydrothermal power system. Hasanien and El-fergany [74] presented a symbiotic organism search algorithm for obtaining the optimal parameters of the frequency controller in the interconnected power system including wind farms. On the other hand, with the rapid growth of RESs, the ESSs such as SMES, electric batteries, fuel cells, and others should be considered within the design of the electricity grid system. Moreover, the ESS has been proposed in a few attempts as a supplementary frequency control (i.e., auxiliary LFC) besides the LFC to support frequency stability of the power systems including RESs. Xie et al. [75] studied the dynamic contribution of the energy storage in the EV battery as an auxiliary LFC technique. El-Saady et al. [76] discussed the effect of SMES technology as a new supplementary AGC scheme in the multi-area power system. Bhatt et al. [77] studied the dynamic contribution of DFIG along with SMES to frequency support of the interconnected power system. Nowadays, utilizing RESs is attracting great attention as a solution to future energy shortages. However, the irregular nature and uncertainties of RESs and random load variations cause many control problems such as large frequency and voltage fluctuations. Hence, this represents a challenge to the power system designers. Therefore, in order to benefit from a maximum capacity of the RESs, new control strategies must be applied in the modern power systems as reported in this chapter. On the other side, the continuous development of computational swarm algorithms represents great motivation for the authors to apply the MSA to find the optimal PID controller parameters in the LFC loops. The MSA is a novel optimization algorithm, which is inspired by the orientation of moths towards moonlight. The available solution of any optimization problem using MSA is performed by the light source position and its fitness is the luminescence intensity of the light source. It was presented by Mohamed et al. [78]. The MSA has some advantages over the other swarm algorithms such as (1) its simplicity and flexibility as it can be applied to different problems without changing the main algorithm structure; (2) ability on avoiding the trap in local minima; (3) achieving fast convergence characteristics; and (4) its lowest control parameters to be adjusted. Therefore, this chapter uses the proposed algorithm to optimally design the PID controller in the LFC loops of the EPS considering HWPP and the dynamic contribution of SMES technology as an auxiliary LFC. The main contribution of this work, a new coordination of the secondary frequency control (i.e., LFC), and the SMES technology (i.e., auxiliary LFC) using a new optimal PID controller-based MSA in the EPS concerning HWPP have been proposed to enhance the frequency stability under the impact of different load profiles, random load variation, wind power fluctuations, and system uncertainties. Also, a comparative study of different control strategies has been carried out to validate the effectiveness of the proposed coordination regarding the peak undershoot, peak overshoot, and settling time. The targeted power system in this study consists of seven strongly tied zones. It is represented as a single area, which comprises dynamics of three subsystems (reheat, non-reheat, and hydropower plants) in addition to integrating wind farms. Further, inherent nonlinearities, GDB, and GRCs of reheat and non-reheat turbine are considered.

2.2 System Configuration and Modeling

2.2.1 Dynamic Model of the EPS

The power system presented in this study is a real power system in Egypt. It is divided into seven strongly tied zones, which are Cairo, Middle Egypt, Upper Egypt, East El-Delta, El-Canal, West El-Delta, and Alexandria as shown in Fig. 2.1. Each zone comprises several power plants (non-reheat, reheat, and hydropower plants or a combination of each). The EPS has almost 180 power plants; moreover, it is classified into three categories: (a) non-reheat power plants represented by gas turbine power plants and a few numbers of steam power plants; (b) reheat power plants mainly

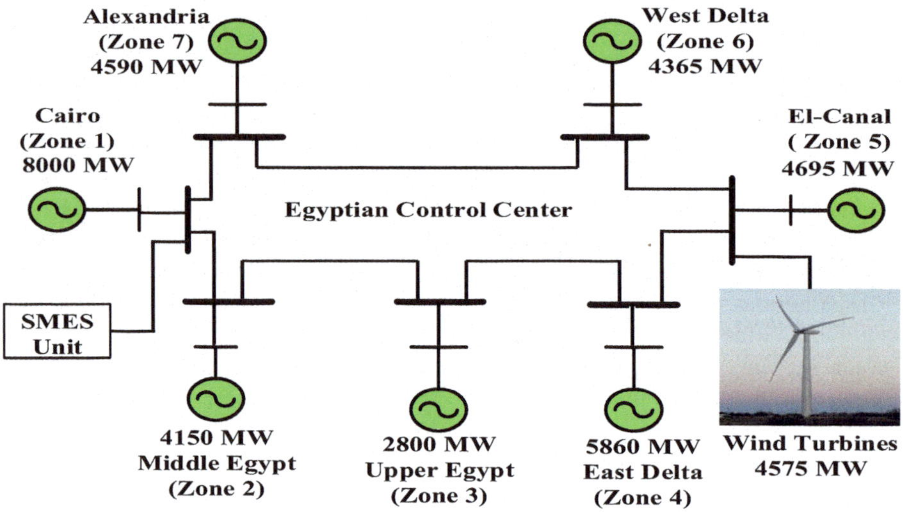

Fig. 2.1 Typical single-line diagram of the EPS

represented by thermal power plants or combined cycle power plants; (c) hydropower plants such as High Dam in Aswan city. Furthermore, the EPS includes RESs, i.e., wind turbines and solar power plants, which contribute almost 3% of the installed capacity. On the other hand, by the year 2020, the Egyptian Electricity Holding Company (EEHC) aims to cover 20% of the electric energy demand from RESs. About 12% of the 20% target is anticipated from wind power and the remaining 8% from other RESs, such as solar energy, e.g., PV power plant, and concentrated solar power plant [79]. The total generation capacity and peak loads are 38,000 and 29,000 MW, respectively, according to the annual report of the EEHC in 2016 [79]. The base of the system frequency is 50 Hz, while the power base is 38,000 MW. Therefore, this chapter considers the EPS with consideration of the effect of HWPP (as future planning of EPS), which includes the proposed coordination of LFC and modified control signal to SMES using a new optimal PID controller-based MSA technique as shown in Fig. 2.2. The National Energy Control Center (NECC) in Egypt has advanced a dynamic model of Egyptian LFC in [80]. Moreover, this model was rebuilt using MATLAB/Simulink with some manipulation, which includes the effect of HWPP, and the dynamic contribution of SMES coordinated with optimal LFC as shown in Fig. 2.3. The NECC in Egypt estimates the system parameter values, which are used in the dynamic model of the EPS as given in Table 2.1.

2.2.2 Mathematical Model of the EPS

The real hybrid power system (e.g., the EPS) considered in this study is the eighth-order linearized multisource power system considering HWPP and SMES technology. The frequency deviation of the studied power system considering the effect of

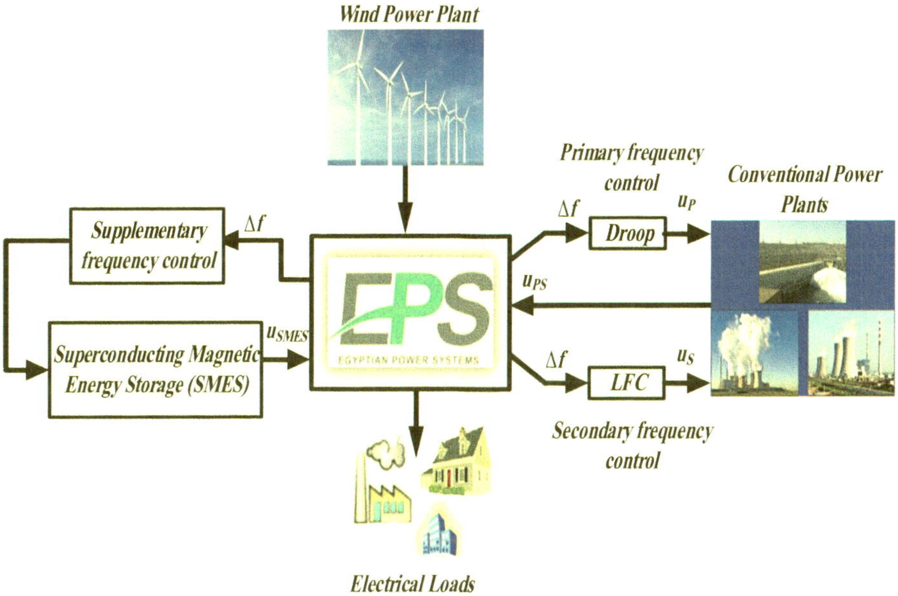

Fig. 2.2 A simplified model of the EPS considering HWPP

the primary control loop (i.e., governor action), a secondary control loop (i.e., LFC), and SMES controller (i.e., supplementary LFC) can be obtained as

$$\Delta f = \frac{1}{2H_{\text{EPS}}s + D_{\text{EPS}}} \times \left(\Delta P_{\text{Non-Reh.}} + \Delta P_{\text{Reh.}} + \Delta P_{\text{Hydro}} + \Delta P_{\text{WT}} \pm \Delta P_{\text{SMES}} - \Delta P_{\text{L}}\right) \tag{2.1}$$

where D_{EPS} is system damping coefficient of the EPS, H_{EPS} is equivalent inertia constant of the EPS, $\Delta P_{\text{Non - Reh.}}$ is power deviation of non-reheat power plant, $\Delta P_{\text{Reh.}}$ is power deviation of reheat power plant, ΔP_{Hydro} is power deviation of hydraulic power plant, ΔP_{SMES} is active power deviation of SMES unit, ΔP_{WT} is wind power variation, and ΔP_{L} is load power variation. Figure 2.3 shows a block diagram of transfer functions describing the different subsystems of the studied nonlinear power system, where the blocks are

1. The non-reheat power plants (e.g., sub-area 1):

$$\Delta P_{\text{Non-Reh.}} = \frac{P_{n1}}{T_1 S + 1} * \left(\frac{-1}{R_1} * \Delta f - \Delta P_c\right) \tag{2.2}$$

2. The reheat power plant (e.g., sub-area 2):

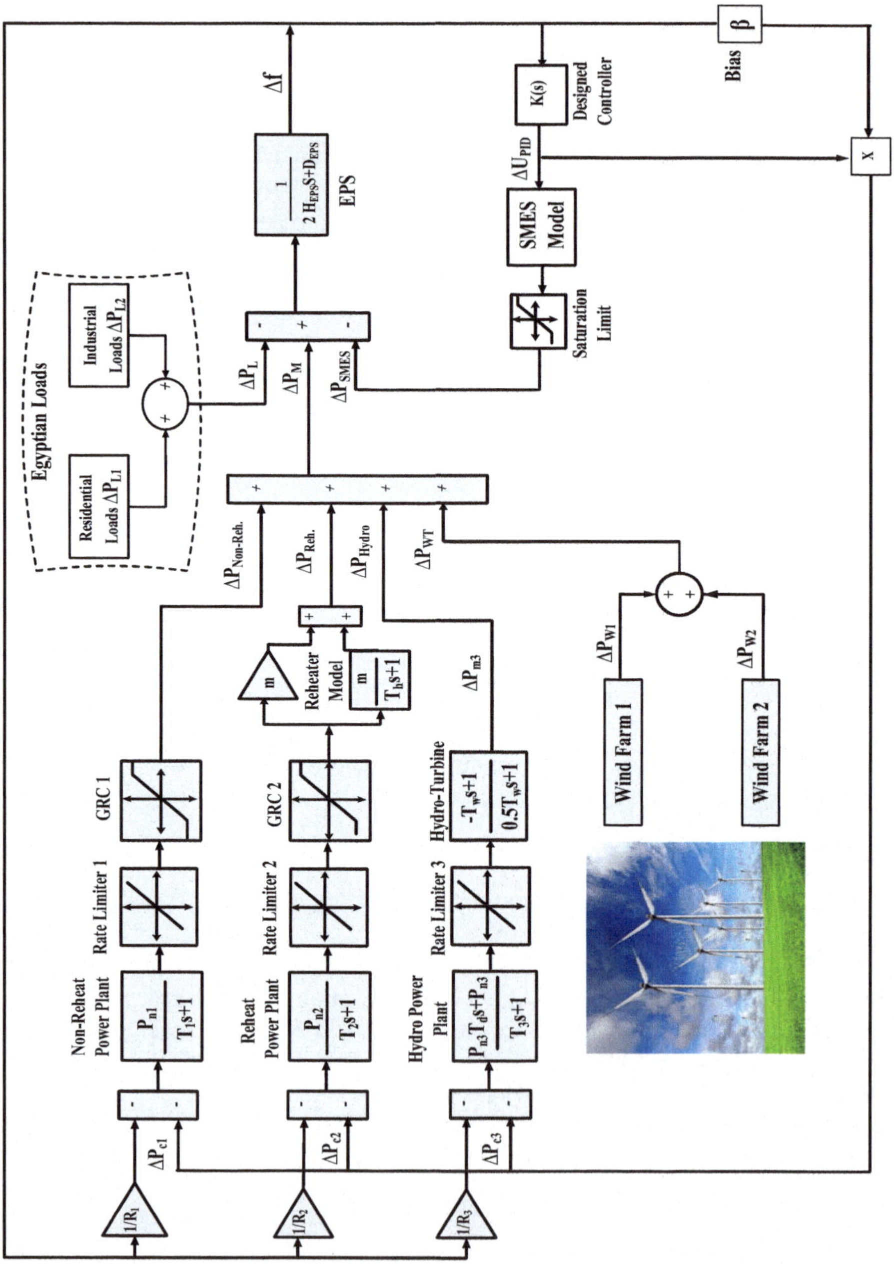

Fig. 2.3 A nonlinear model of the EPS considering HWPP with the proposed control strategy

Table 2.1 System parameters of the EPS [81]

Parameter	Value	Parameter	Value	Parameter	Value
D_{EPS}	0.028	m	0.5	P_{n3}	0.1364
T_1	0.4	R_1	2.5	$V_{U,1}$	0.05
T_2	0.4	R_2	2.5	$V_{L,1}$	−0.05
T_3	90	R_3	1.0	$V_{U,2}$	0.05
T_d	5	H_{EPS}	5.7096	$V_{L,2}$	−0.05
T_h	6	P_{n1}	0.2529	$V_{U,3}$	0.01
T_w	1.0	P_{n2}	0.6107	$V_{L,3}$	−0.01
GRC1	0.2	GRC2	0.1		

$$\Delta P_{g2} = \frac{P_{n2}}{T_2 S + 1} * \left(\frac{-1}{R_2} * \Delta f - \Delta P_c\right). \tag{2.3}$$

$$\Delta P_{\mathrm{Reh.}} = \left(m + \frac{m}{T_h S + 1}\right) * \Delta P_{g2} \tag{2.4}$$

3. The hydropower plants (e.g., sub-area 3):

$$\Delta P_{g3} = \frac{P_{n3} T_{\mathrm{d}} S + P_{n3}}{T_3 S + 1} * \left(\frac{-1}{R_3} * \Delta f - \Delta P_c\right). \tag{2.5}$$

$$\Delta P_{\mathrm{Hydro.}} = \left(\frac{-T_{\mathrm{w}} S + 1}{0.5 * T_{\mathrm{w}} S + 1}\right) * \Delta P_{g3} \tag{2.6}$$

4. The wind turbine model:

$$\Delta P_{\mathrm{WT}} = \frac{1}{T_{\mathrm{WT}} S + 1} * (\Delta P_{\mathrm{Wind}}) \tag{2.7}$$

5. The SMES model:

$$\Delta P_{\mathrm{SMES}} = \frac{K_{\mathrm{SMES}}}{T_{\mathrm{SMES}} S + 1} * (\Delta f) \tag{2.8}$$

In this study, the wind power variation (ΔP_{WT}) and load power variation (ΔP_{L}) are considered as power system disturbance signals. However, aggregated wind power farms are modeled as high, and low-fluctuated wind power as load model. Utilizing Eqs. (2.1)–(2.8) and the block diagram of Fig. 2.3, the dynamic equations of the studied hybrid power system can be derived and written in the state variable form as follows:

$$\dot{X} = \mathrm{AX} + \mathrm{BU} + \mathrm{EW} \tag{2.9}$$

$$Y = \mathrm{CX} + \mathrm{DU} + \mathrm{FW} \tag{2.10}$$

Hence, the complete state-space equations for the EPS considering HWPP and the effect of the dynamic contribution of SMES technology can be obtained as in Eqs. (2.11) and (2.12):

$$\dot{X} = \begin{bmatrix} -\frac{D_{\mathrm{EPS}}}{2H_{\mathrm{EPS}}} & \frac{1}{2H_{\mathrm{EPS}}} & \frac{1}{2H_{\mathrm{EPS}}} & 0 & \frac{1}{2H_{\mathrm{EPS}}} & 0 & \frac{1}{2H_{\mathrm{EPS}}} & \frac{1}{2H_{\mathrm{EPS}}} \\ -a_1 & -\frac{1}{T_1} & 0 & 0 & 0 & 0 & 0 & 0 \\ -ma_2 & 0 & -\frac{1}{T_h} & b_1 & 0 & 0 & 0 & 0 \\ -a_2 & 0 & 0 & -\frac{1}{T_2} & 0 & 0 & 0 & 0 \\ (-2b_2D_{\mathrm{EPS}}+2a_3) & 2b_2 & 2b_2 & 0 & \left(2b_2-\frac{2}{T_{\mathrm{w}}}\right) & \left(\frac{2}{T_{\mathrm{w}}}+\frac{2}{T_3}\right) & 0 & 0 \\ (b_2D_{\mathrm{EPS}}-a_3) & -b_2 & -b_2 & 0 & -b_2 & -\frac{1}{T_3} & 0 & 0 \\ 0 & 0 & 0 & 0 & 0 & 0 & -\frac{1}{T_{\mathrm{WT}}} & 0 \\ \frac{K_{\mathrm{SMES}}}{T_{\mathrm{SMES}}} & 0 & 0 & 0 & 0 & 0 & 0 & -\frac{1}{T_{\mathrm{SMES}}} \end{bmatrix}$$

$$\times \begin{bmatrix} \Delta f \\ \Delta P_{\mathrm{Non-Reh}} \\ \Delta p_{\mathrm{Reh.}} \\ \Delta p_{g2} \\ \Delta P_{\mathrm{Hydro.}} \\ \Delta P_{g3} \\ \Delta P_{\mathrm{WT}} \\ \Delta P_{\mathrm{SMES}} \end{bmatrix} + \begin{bmatrix} 0 \\ -\frac{P_{n1}}{T_1} \\ -\frac{m*P_{n2}}{T_2} \\ -\frac{P_{n2}}{T_2} \\ \frac{2*P_{n3}}{T_3} \\ -\frac{P_{n3}}{T_3} \\ 0 \\ 0 \end{bmatrix} [\Delta P_c] + \begin{bmatrix} 0 & -\frac{1}{2H_{\mathrm{EPS}}} \\ 0 & 0 \\ 0 & 0 \\ 0 & 0 \\ 0 & -2b_2 \\ 0 & b_2 \\ -\frac{1}{T_{\mathrm{WT}}} & 0 \\ 0 & 0 \end{bmatrix} \begin{bmatrix} \Delta P_{\mathrm{Wind}} \\ \Delta P_{\mathrm{L}} \end{bmatrix} \tag{2.11}$$

$$Y = [\beta\ 0\ 0\ 0\ 0\ 0\ 0\ 0] \begin{bmatrix} \Delta f \\ \Delta P_{\text{Non-Reh}} \\ \Delta p_{\text{Reh.}} \\ \Delta p_{g2} \\ \Delta P_{\text{Hydro.}} \\ \Delta P_{g3} \\ \Delta P_{\text{WT}} \\ \Delta P_{\text{SMES}} \end{bmatrix} + [0][\Delta P_c] + [00] \begin{bmatrix} \Delta P_{\text{Wind}} \\ \Delta P_{\text{L}} \end{bmatrix} \tag{2.12}$$

The constants of matrices are $a_1 = \frac{P_{n1}}{T_1 R_1}$, $a_2 = \frac{P_{n2}}{T_2 R_2}$, $a_3 = \frac{P_{n3}}{T_3 R_3}$, $a_4 = \frac{T_{\text{d}}}{2H_{\text{EPS}}}$, $b_1 = \frac{2m}{T_h} - \frac{m}{T_2}$, $b_2 = a_3 a_4$.

2.2.3 Wind Power Generation System

Figure 2.4 shows the model of the wind power generation system (WPGS) for frequency control [45, 81]. In this model, the wind speed is multiplied by the random speed fluctuation, which is derived from the white noise block in MATLAB/Simulink to estimate the random wind output power fluctuations. The WPGS model can be described by the following equations [74]:

The wind turbine output power is calculated as follows:

$$P_{\text{Wind}} = \frac{1}{2} \rho A_{\text{T}} V_{\text{W}}^3 C_{\text{P}}(\lambda, \beta) \tag{2.13}$$

where ρ is the air density (kg/m^3), A_{T} is the rotor swept area (m^2), V_{W} is the rated wind speed (m/s), and C_{P} represents the power coefficient of the rotor blades. C_{P} is defined below in terms of turbine coefficients C_1 to C_7:

$$C_{\text{P}}(\lambda, \beta) = C_1 * \left(\frac{C_2}{\lambda_{\text{I}}} - C_3 \beta - C_4 \beta^2 - C_5 \right) * e^{\frac{-C_6}{\lambda_I}} + C_7 \lambda_{\text{T}} \tag{2.14}$$

where β is the pitch angle and λ_{T} corresponds to the optimal tip-speed ratio (TSR) defined by the following:

$$\lambda_{\text{T}} = \lambda_{\text{T}}^{op} = \frac{\omega_{\text{T}} * r_{\text{T}}}{V_{\text{W}}} \tag{2.15}$$

Variable speed wind turbines operate at an optimal TSR value during all wind speed conditions, and r_{T} is the rotor radius. In Eq. (2.14), λ_{I} is intermittent TSR and is related to λ_{T} and β as demonstrated below:

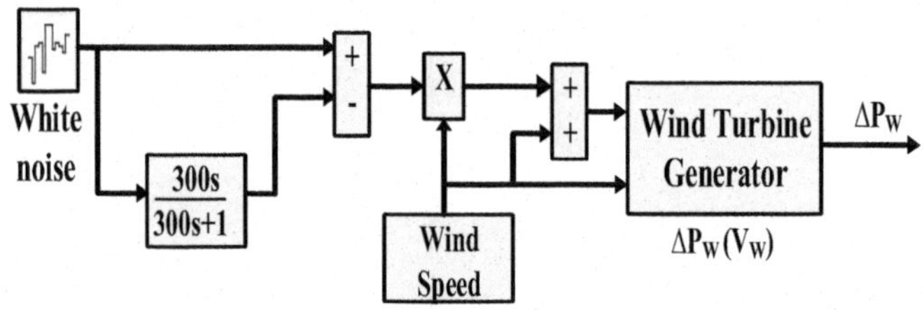

Fig. 2.4 The model of WPGS using MATLAB/Simulink [81]

$$\frac{1}{\lambda_{\mathrm{I}}} = \frac{1}{\lambda_{\mathrm{T}} + 0.08\beta} - \frac{0.035}{\beta^3 + 1} \tag{2.16}$$

This study considers the effect of merging HWPP on the system frequency response of a real hybrid power system. In this study, the EPS includes aggregated wind turbine models with different power rates, which are 1300 wind turbine units of 750 kW for each unit (i.e., wind farm 1), and 1200 wind turbine units of 3.0 MW for each unit (i.e., wind farm 2) besides the conventional generation units. The EPS is tested in the presence of low-fluctuated wind power (i.e., wind farm 1) and high-fluctuated wind power (i.e., wind farm 2) as shown in Fig. 2.5. The parameter values of each wind turbine model are presented in Table 2.2 and Table 2.3.

2.2.4 Modeling of Power System Loads

In this study, the input power fluctuation of the EPS is estimated by considering the deviation from the initial load value as shown in Fig. 2.6. The standard deviation is multiplied by the random output fluctuations derived from the white noise block in MATLAB/Simulink in order to simulate the real-time random power fluctuations on the load profile. However, low random load change (i.e., residential load) and high random load change (i.e., industrial load) are tested on the EPS as shown in Fig. 2.7. The load deviation is simulated close to an actual load change by the following function [45, 74]:

$$\Delta P_{\mathrm{L}} = 0.6\sqrt{P_{\mathrm{Load}}} \tag{2.17}$$

2.2.5 Modeling of SMES Technology in LFC

SMES is a device, which can control the active and reactive power simultaneously via storage of the electrical power from the grid. It stores the power in the magnetic coil,

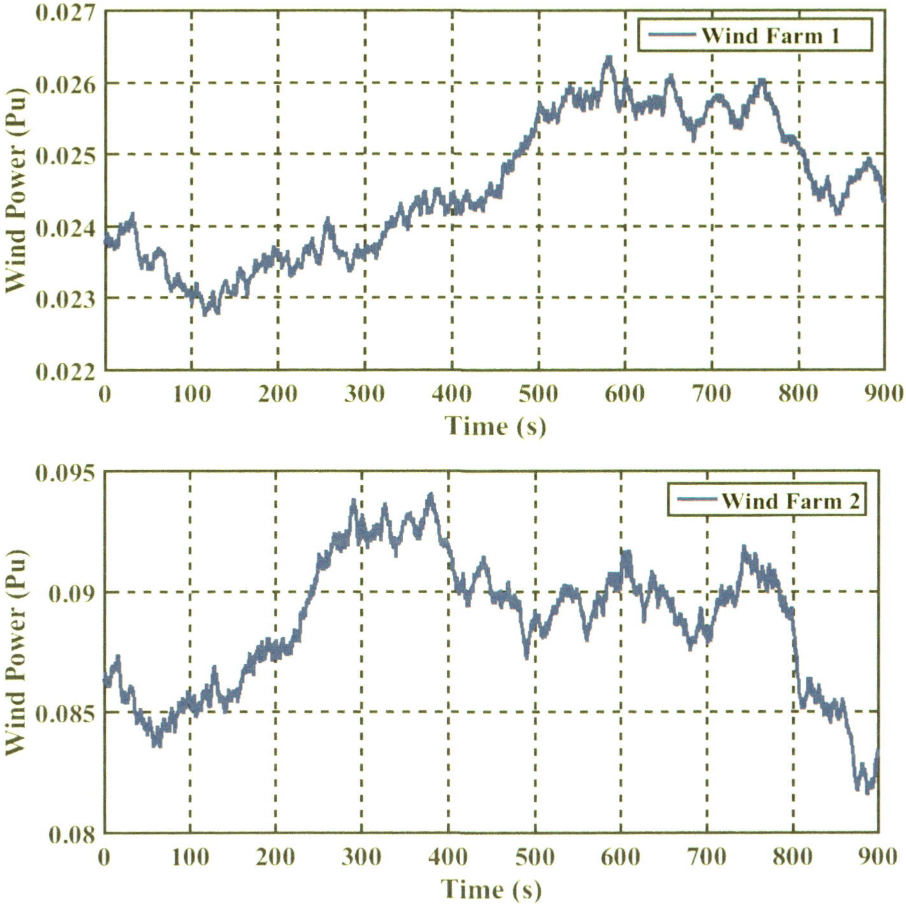

Fig. 2.5 The wind power output profiles of the EPS

Table 2.2 Nominal wind turbine parameters of wind farm 1

Parameter	Value	Parameter	Value
$P_{W,1}$	750 kW	C_2	116
$V_{W,1}$	15 m/s	C_3	0.4
ρ	1.225 kg m^2	C_4	0
A_T	1648 m^2	C_5	5
r_T	22.9 m	C_6	21
n_T	22.5 rpm	C_7	0.1405
C_1	−0.6175		

which is made from a superconducting material with nearly zero loss of energy [24]. Also, it is one of the most effective stabilizers of frequency oscillations. It has fast response, high efficiency, and long lifetime compared to other energy ESSs. This makes it a preferable choice for an energy storage solution in different power systems,

Table 2.3 Nominal wind turbine parameters of wind farm 2

Parameter	Value	Parameter	Value
$P_{W,2}$	3000 kW	C_2	116
$V_{W,2}$	12 m/s	C_3	0.4
ρ	1.225 kg m^2	C_4	0
A_T	5905 m^2	C_5	5
r_T	43.36 m	C_6	21
n_T	22.5 rpm	C_7	0.0192
C_1	0.3915		

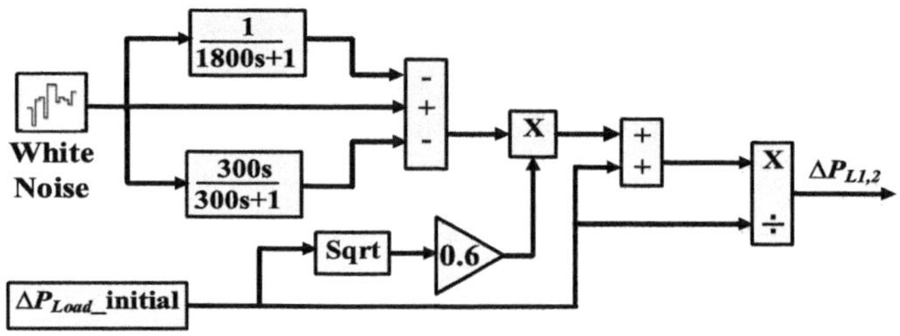

Fig. 2.6 The model of random load using MATLAB/Simulink [81]

which are integrated with RESs. Hence, it has attracted many researchers to study its potential applications in power systems. With the technology improvement and discovering of new superconducting materials, the predicted cost of the SMES unit is likely to drop down to 25% of the existing cost [77]. Recently, the SMES device would penetrate the market and be used for LFC application (i.e., auxiliary LFC) [82]. SMES has the ability to charge and discharge electrical power from/to the grid with a very short time considering the SMES power limits. It has been used in this study as frequency stabilizer (i.e., auxiliary secondary control) for more damping out of the system frequency oscillations, which are generated from the power fluctuations of RESs, random load change, and system parameter variation.

The linearized model representation of SMES such as a control scheme is shown in Fig. 2.8. The SMES device is simulated as a first-order transfer function by a time constant T_{SMES} and SMES variable gain K_{SMES}. The criteria for selecting these two parameters are related to the stability of the system and the response required in dynamics. Moreover, the initial rate of change of frequency (RoCoF) as well as the maximum drop of frequency (i.e., nadir frequency) must be taken into consideration. In this study, the SMES parameters T_{SMES} and K_{SMES} are selected as 0.03 s and 6, respectively, which are obtained using the trial-and-error method. This selection gives good dynamic stability during the transients even when the total system inertia is reduced to 50% of its system. The SMES power as in Eq. (2.8) is obtained by estimating RoCoF as shown below:

$$\text{RoCoF} = \frac{\mathrm{d}(\Delta f)}{\mathrm{d}t} \tag{2.18}$$

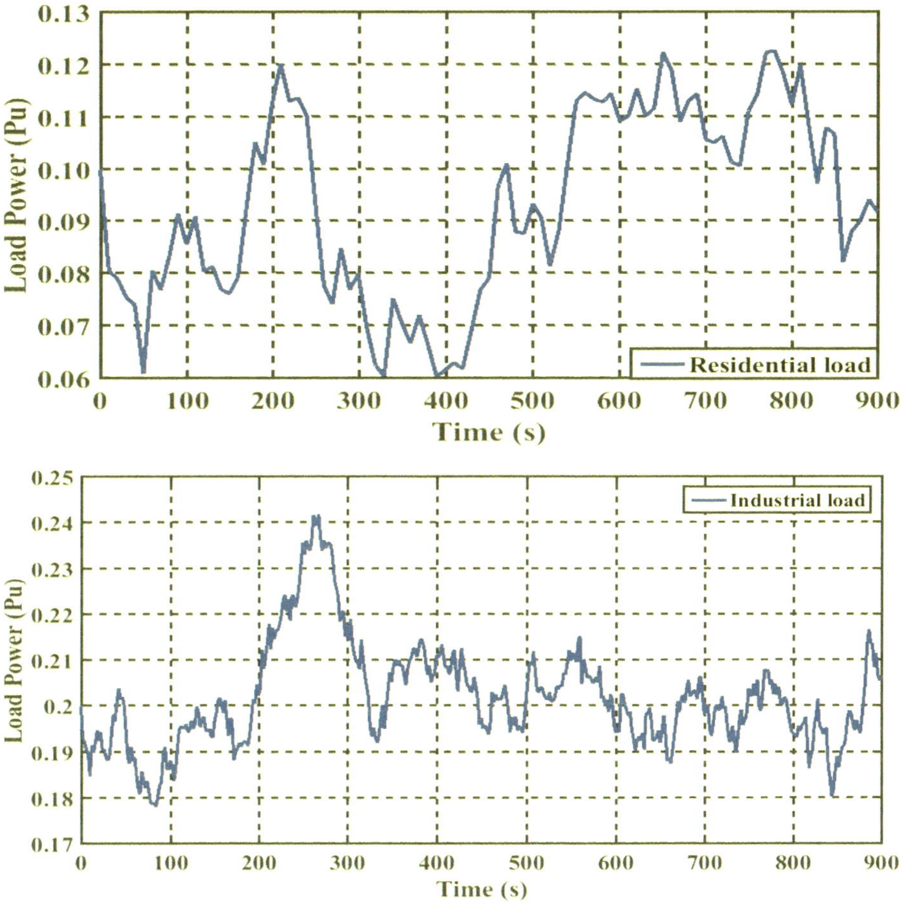

Fig. 2.7 The random load fluctuations of the studied cases

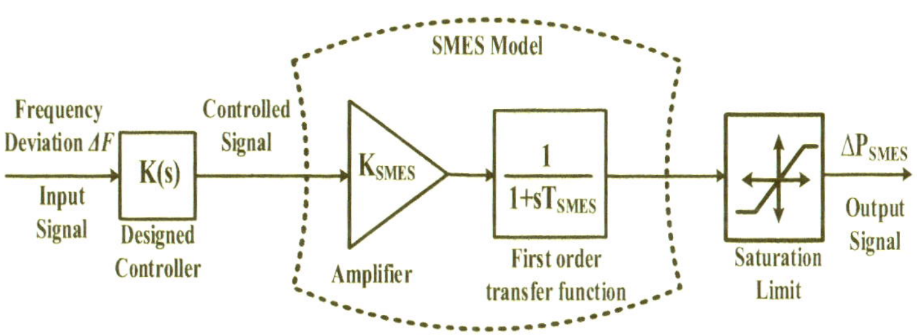

Fig. 2.8 Structure of SMES model as frequency stabilizer with a designed controller

2.3 Control Methodology and Problem Formulation

In this chapter, the proposed coordinated control strategy of secondary frequency control (i.e., LFC) and SMES (i.e., auxiliary LFC) is based on the PID controller in the EPS considering HWPP. The PID controller is composed of three terms (i.e., gains), which are proportional gain K_p, integral gain K_i, and derivative gain K_d. Its transfer function is expressed as follows:

$$G(s) = K_p + \frac{K_i}{s} + K_d s \tag{2.19}$$

A novel intelligent searching method (i.e., the MSA) is used to find the optimal design parameters of the PID controller, which has parameter bounds. In this study, the integral of squared error (ISE) is used as a fitness function, which is the objective function of the proposed optimization technique and can be formulated as follows:

$$\mathrm{ISE} = \int_0^{t\mathrm{sim}} (\Delta f)^2 \mathrm{d}t \tag{2.20}$$

This is subject to bounds of the PID controller parameters as follows:

$$K_{p,i,d}^{\mathrm{Min}} \leq K_{p,i,d} \leq K_{p,i,d}^{\mathrm{Max}} \tag{2.21}$$

where Δf is the frequency deviation of the EPS and t_{sim} is the simulation time to execute one run. The proposed MSA technique is applied in the EPS to obtain the minimum value of objective function through the optimal parameters of the PID controller.

2.4 Moth Swarm Algorithm

In this study, the MSA has been adapted to find the optimal parameters of the PID controller. The MSA technique was presented in [78] as a simple and fast searching intelligent technique. In the proposed MSA, the possible solution of the optimization problem is represented by the position of light source, and the fitness/quality of this solution is considered as luminescence intensity of the light source. These assumptions have been used to approximate the characteristics of the proposed algorithm. Furthermore, the proposed method consists of three main groups:

1. The first group is called pathfinders, which are considered as a small group of moths over the available space of the optimization. The main target of this group is to guide the locomotion of the main swarm by discriminating the best positions as light sources.

2. The prospectors group, which has a tendency to expatiate in a nonuniform spiral path within the section of the light sources determined by the pathfinders.
3. The onlookers group: This group of moths moves directly to the global solution, which has been acquired by the prospectors.

With MSA, different optimization operators are used to mimic a set of behavioral patterns of moths in nature, which allows for the flexible and powerful optimizer. Hence, a new dynamic selection strategy of crossover points is proposed based on population diversity to handle the difference vector Lévy mutation to force the MSA to jump out of stagnation and enhance its exploration ability. In addition, a spiral motion, adaptive Gaussian walks, and a novel associative learning mechanism with immediate memory are implemented to exploit the promising areas in the search space [78]. The new position of *i*thonlooker moth can be expressed mathematically as follows:

$$x_i^{t+1} = x_i^t + \varepsilon_1 + \left[\varepsilon_2 \times \text{best}_g^t - \varepsilon_3 \times x_i^t\right] \forall i\{1, 2, \ldots, n_G\} \tag{2.22}$$

where

$$n_G = \text{round}(n_o/2).$$

The best previous position of any onlooker is recorded, and this value is called best_p. Another best value that is tracked by a global version of the MSA is the overall best value and is called best_g. The updating equation of this type can be completed in the form

$$x_i^{t+1} = x_i^t + 0.001 \cdot G\left[x_i^t - x_i^{\min}, x_i^{\max} - x_i^t\right] + \left(1 - \frac{g}{G}\right) \cdot r_1 \cdot \left(\text{best}_p^t - x_i^t\right) + 2g/G \cdot r_2 \left(\text{best}_g^t - x_i^t\right) \tag{2.23}$$

where $i \in \{1, 2, \ldots, (n_o - n_G)\}$, t is the number of iterations, $\left(\frac{2g}{G}\right)$ is the social factor, $\left(1 - \frac{g}{G}\right)$ is the cognitive factor, and best_p is a light source randomly selected from the new group of pathfinders based on the probability value of its consistent solution. Therefore, in this study, the main objective of the MSA is to minimize the objective function as in Eq. (2.20) through finding the optimal parameters of the PID controller (i.e., K_p, K_i, and K_d). The code of the MSA is performed using MATLAB software, and the steps of the proposed algorithm can be concluded as shown in Table 2.4. The performance of the MSA-PID in searching the PID controller parameters of the LFC loops in the EPS has been validated by using the characteristics of the MSA as given in Table 2.5. These optimal characteristics are selected by the trial-and-error method, which is the most commonly used with optimization algorithms. Therefore, the optimal parameters of the PID controller-based MSA under the system operation condition without RESs are given in Table 2.6. The parameters of the PID controller lie in the range [0, 100] according to the trials.

Table 2.4 Pseudo-code of the proposed MSA

Calculate the swarm finesses and identify the type of each moth
While $t <$ max number of iterations T For each light source: • Identify the crossover points $c_p \subset \{c_1, c_2, \ldots, c_{n_c}\}$. • Generate Lévy flight samples L_p. • Create/mutate sub-trail vector $\overrightarrow{v_p^t}$. • Construct the completed trail solution $\overrightarrow{V_p^t}$. • Select the artificial light sources $\overrightarrow{x_p^{t+1}}$. End of reconnaissance.
Calculate the probability values P
For each prospector moth x_i: • Update the position of prospector moth. • Calculate the fitness of prospector. End of transverse orientation.
Define the new light sources and moonlight
For each onlooker moth (worse group): • Update the position according to its type. If ($i \in n_G$), Generate Gaussian walk steps ε_1, ε_2, and ε_3. Move the onlooker position with Gaussian walks x_i^{t+1}. Else, Drift the onlooker moth using the associative learning operators and immediate memory. End if. • Calculate the fitness of onlooker moth. End of celestial navigation.
Identify the new light sources, moonlight, and type of each moth. End while. Print global best solution (moonlight).

2.5 Simulation Study

In this study, the proposed coordination performance of LFC and SMES using a new optimal PID controller-based MSA is compared with both the optimal LFC-based MSA with/without the effect of conventional SMES (which is without modifying the input control signal) for the realistic hybrid power system (i.e., the EPS) in the presence of HWPP considering the effect of nonlinearity. The EPS comprises steam power plants (reheat and non-reheat turbine), gas power stations (non-reheat turbine), hydropower plants, wind power plants, and SMES technology. Three subsystems (non-reheat, reheat, and hydropower plants) are given for the EPS with inherent nonlinearities, which are speed governor backlash and GRCs of power plants. Backlash is defined as the total magnitude of sustained speed change. All speed governors have a backlash, which is important for LFC in the presence of disturbances. The GRCs limit the generation rate of the output power which is given as 0.2 pu MW/min and 0.1 pu MW/min for non-reheat and reheat turbines, respectively. Moreover, the actual GRC of a hydropower plant is about 0.5 pu MW/min,

Table 2.5 The control parameters of MSA

Parameter	Value
Maximum iteration	50
Search agent size	50
Pathfinders number	10
Problem dimension	3
Average simulation time t_{sim} (s)	229.5

Table 2.6 Optimal values of the MSA-based PID controller

Parameter	K_p	K_i	K_d
Value	71.2532	5.9055	6.10758

which is higher than the generation rate corresponding to any practical disturbance, and hence it will be neglected [8].

2.6 Simulation Results and Discussion

The model of the targeted power system (i.e., the EPS) including HWPP with inherent nonlinearities is built using MATLAB/Simulink model, which takes the GRCs of different generation sources into account. The code of the MSA as an m-file is interfaced with the EPS model to execute the optimization process. The simulation program is carried out on a PC with an Intel® Core i5–2.6 GHz, 8.0 GB RAM. Analysis of different control strategies has been implemented on the EPS. Many researches assume that the load profile is a step change disturbance, which is represented by the forced outage of generation unit or sudden switch off of a massive load. However, in fact, load disturbances are complex and of random nature. The actual load disturbance can be represented by several types of disturbance besides step change disturbance. Therefore, this study applied some random load variations, which are low random load change (residential load) and high random load change (industrial load) on the EPS in the presence of low and high wind power fluctuation models for wind farms 1 and 2, respectively. Two scenarios are applied to the studied system and the simulation time of each one is 15 min as follows:

Scenario A: In this scenario, the EPS considering HWPP as shown in Fig. 2.3 is considered as the test system to confirm the robustness and effectiveness of the proposed coordination. The proposed coordination of LFC and SMES technology using the designed PID controller, whose values are listed in Table 2.6, is tested by implementing the high-fluctuated wind power (i.e., wind farm 2) at 500 s. In addition, the other disturbance sources (i.e., low-fluctuated wind power, residential load, and industrial load) are connected to the EPS at $t = 0$. Figure 2.9 shows the frequency deviation of three different control strategies for the EPS considering the HWPP. It is clear that the dynamic contribution of conventional SMES improves the frequency response and reduces the transient frequency deviation compared with the Egyptian LFC system without SMES. Moreover, in the case of no SMES controller, the frequency fluctuates more with high deviation. On the other hand, the proposed control strategy and new coordination of LFC and SMES using a new optimal PID controller-based MSA give superior performance and more reduction of the frequency excursions (Δf) and the wind power fluctuations than other strategies. The performance specifications, maximum overshoot (MOS), maximum undershoot (MUS), and maximum settling time (T_S) of the studied system, with different control strategies of SMES under the study of scenario A during the whole period of simulation (15 min) have been compared in Table 2.7.

Scenario B: In this scenario, the effectiveness and robustness of the proposed coordination are evaluated under system parameter variations (i.e., system uncertainty). Moreover, the wind farms (i.e., high- and low-fluctuated wind farms) and different random loads (i.e., residential and industrial loads) are applied to the EPS under the assumed multiple operating conditions in Table 2.8. Hence, as it is known, the PID controller is very sensitive to the system uncertainty and nonlinearity, which represents the main demerit of this controller [83]. Therefore, the PID controller is optimally designed using the MSA under these system operation conditions. Table 2.9 lists the optimal values of the PID controller-based MSA for these system operation conditions. This scenario is divided into three sub-scenarios (i.e., B1, B2, and B3); the purpose of these sub-scenarios is to test the robustness of the proposed coordination control strategy against system inertia variations.

Scenario B1: In this case, the EPS is considered as the test system with high system inertia (100% of default system inertia) considering the multiple operating conditions of wind and load variations as shown in Table 2.8. The frequency deviation of the EPS with different control strategies of SMES is illustrated in Fig. 2.10. From Fig. 2.10, it has been noticed that the frequency deviation of the EPS considering HWPP with the proposed coordination is about ±0.0081 Hz, while the EPS with/without the dynamic contribution of convention SMES is ±0.0129 and ±0.0256 Hz, respectively. Therefore, the system response using the proposed coordination is faster, has a lower steady-state error, and is better damped than other control strategies.

Scenario B2: In this case, the EPS considering HWPP is tested under the condition of 75% of default system inertia with multiple operating conditions of wind and load variations as shown in Table 2.8. The frequency deviation of the three control strategies for the studied system considering the low and high fluctuation of wind power and load with high penetration of RESs is investigated in Fig. 2.11. As per this figure, in the case of no SMES controller, the frequency deviation is about ±0.033 Hz, while the EPS with the dynamic contribution of conventional SMES gives the frequency deviation of about ±0.016 Hz when the industrial load is connected to the EPS at 300 s. Compared to the EPS considering HWPP with/without SMES controller, the proposed coordination control strategy can provide smooth and secure frequency performance during the multiple operation conditions of the EPS

Scenario B3: In this case, the robustness of the proposed coordination of LFC and SMES technology using a new optimal PID-based MSA is tested in the extreme scenario. In this case, the system inertia is decreased to half of the default system

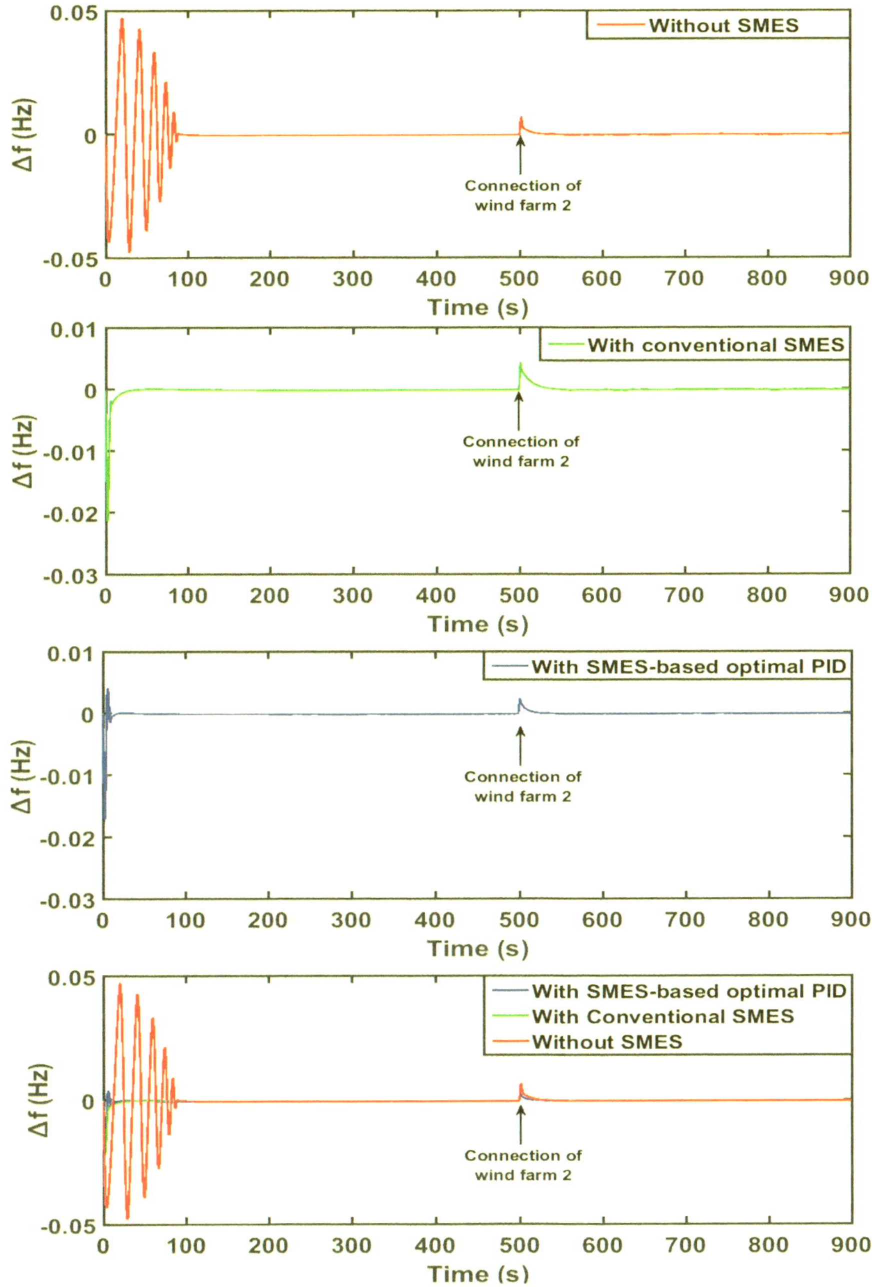

Fig. 2.9 The frequency deviation of the EPS for scenario A

Table 2.7 The performance specification of the studied system for scenario A

Scenario A	With SMES-based PID	With SMES	Without SMES
MUS (pu)	0.0173	0.021	0.0466
MOS (pu)	0.0042	0.0044	0.0476
T_S (s)	19	38	90

Table 2.8 Multiple operating conditions of the EPS considering HWPP for scenario B

Disturbance source	Starting time (s)	Stopping time (s)	Rate power (pu)
Industrial load	300 s	–	0.24
Residential load	Initial	700 s	0.12
Wind farm 1	Initial	–	0.026
Wind farm 2	500 s	–	0.094

Table 2.9 Optimal values of the MSA-based PID controller for the EPS with the system uncertainty effect

Parameter	K_p	K_i	K_d
Value	55.1085	6.0908	0.6140

inertia (50% of default system inertia). This can happen in case of integration of high level of wind power, in which the power electronic interface-based RESs will reduce the overall system inertia and cause to fluctuate the system frequency. Figure 2.12 shows the frequency deviation of the EPS considering HWPP under the condition of 50% of default system inertia with multiple operating conditions of wind and load variations as shown in Table 2.8. As shown in Fig. 2.12, the simulation results concluded that the frequency fluctuates more with high deviation, whereas the studied system with no SMES controller oscillates to such an extent that it is not acceptable. The proposed coordination control strategy can reduce and stabilize the frequency deviation of the EPS and give robustness to the system subjected to uncertainties over the conventional SMES. The performance specifications like MOS, MUS, and T_S of the studied system with different control strategies of SMES under all studied cases of scenario B during the whole period of simulation (15 min) have been compared in Table 2.10.

2.7 Conclusion

This chapter has proposed new coordination of LFC and SMES technology using a new PID controller-based MSA to enhance the frequency stability in the realistic hybrid power system (e.g., the EPS) concerning HWPP as future planning. The EPS is decomposed into three subsystems: non-reheat, reheat, and hydropower plants with inherent nonlinearities considering RESs. A recently developed hybrid algorithm, namely MSA, has been applied to optimize the PID controller parameters with

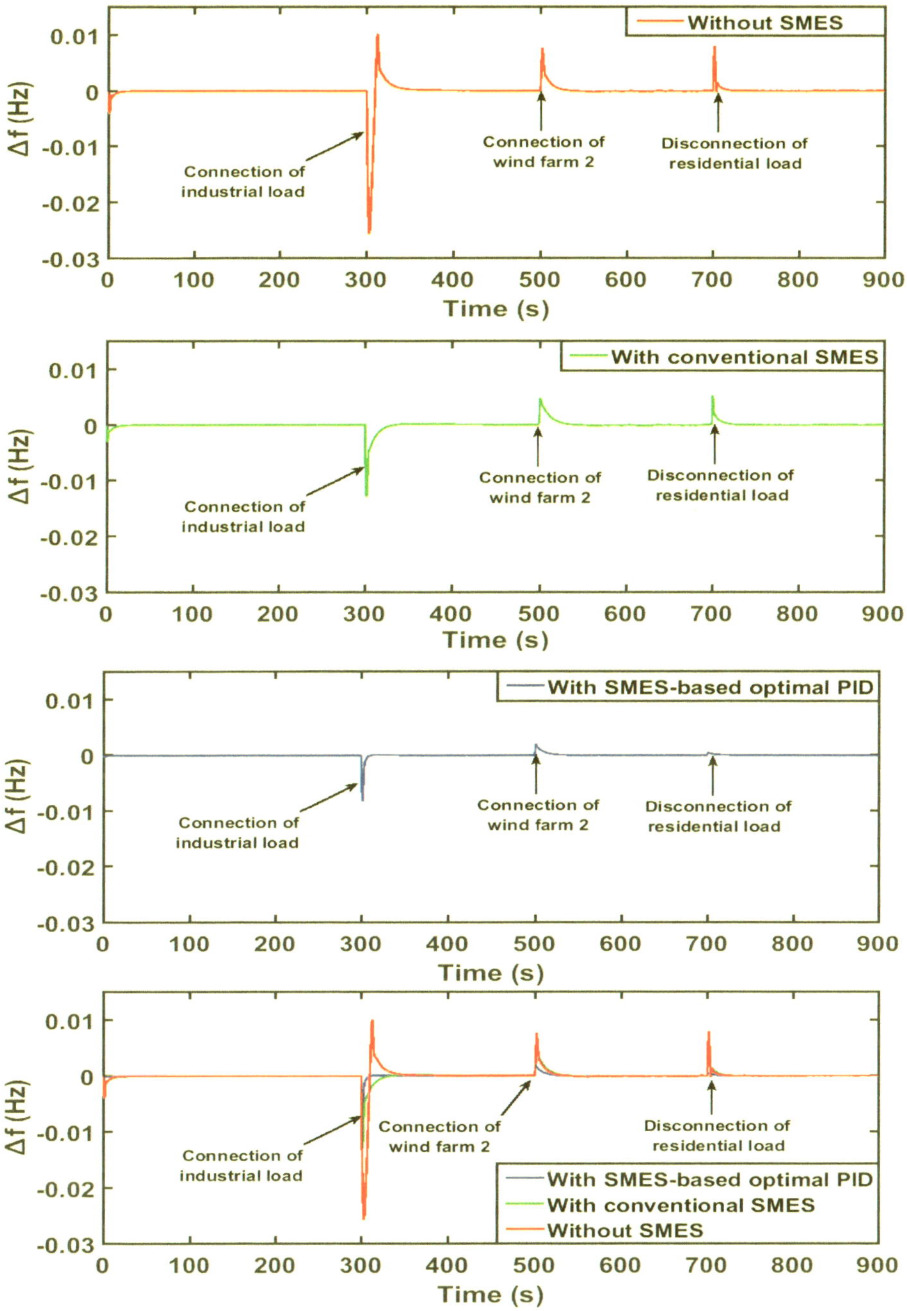

Fig. 2.10 The frequency deviation of the EPS for scenario B1 (i.e., 100% of default system inertia)

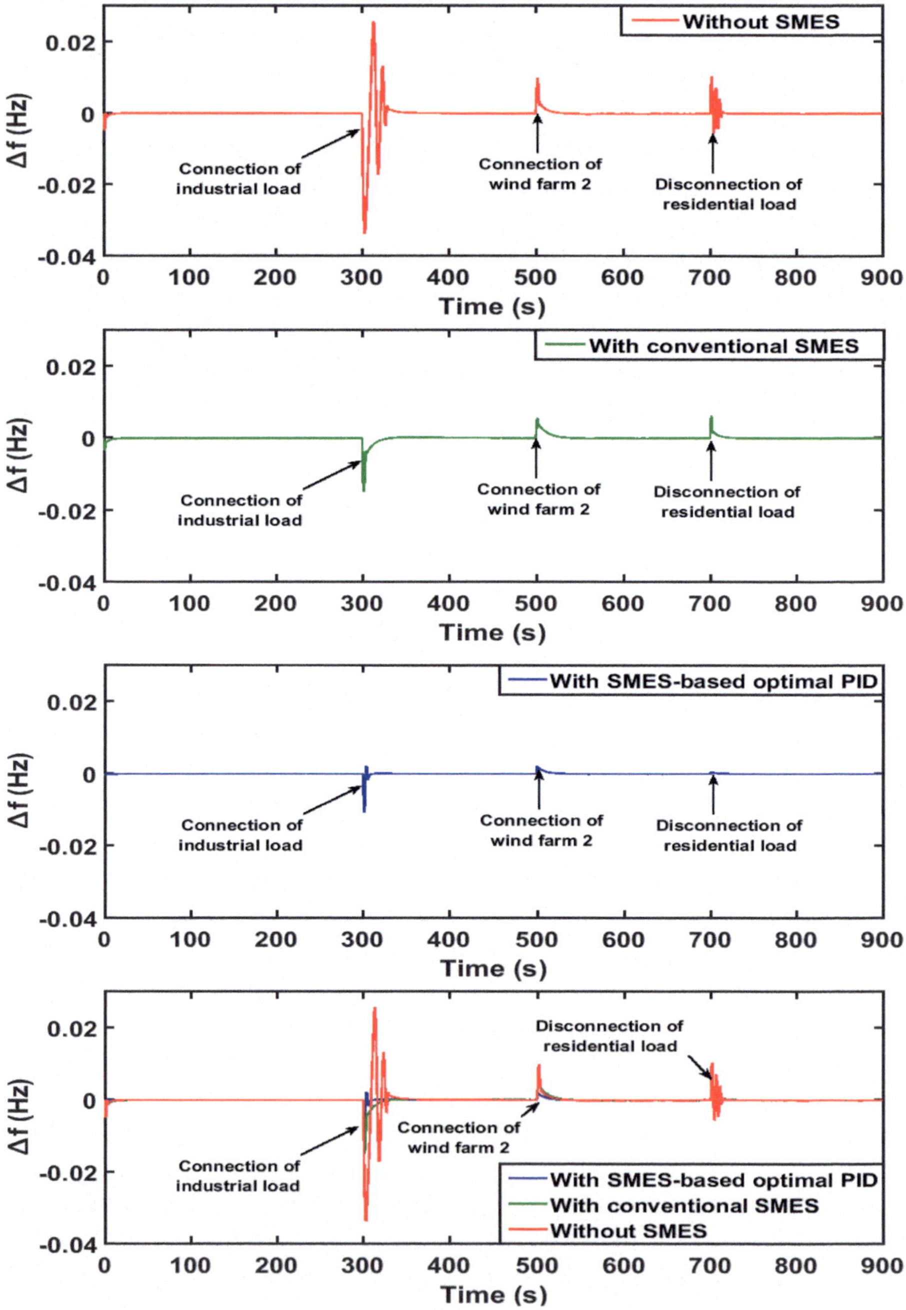

Fig. 2.11 The frequency deviation of the EPS for scenario B2 (i.e., 75% of default system inertia)

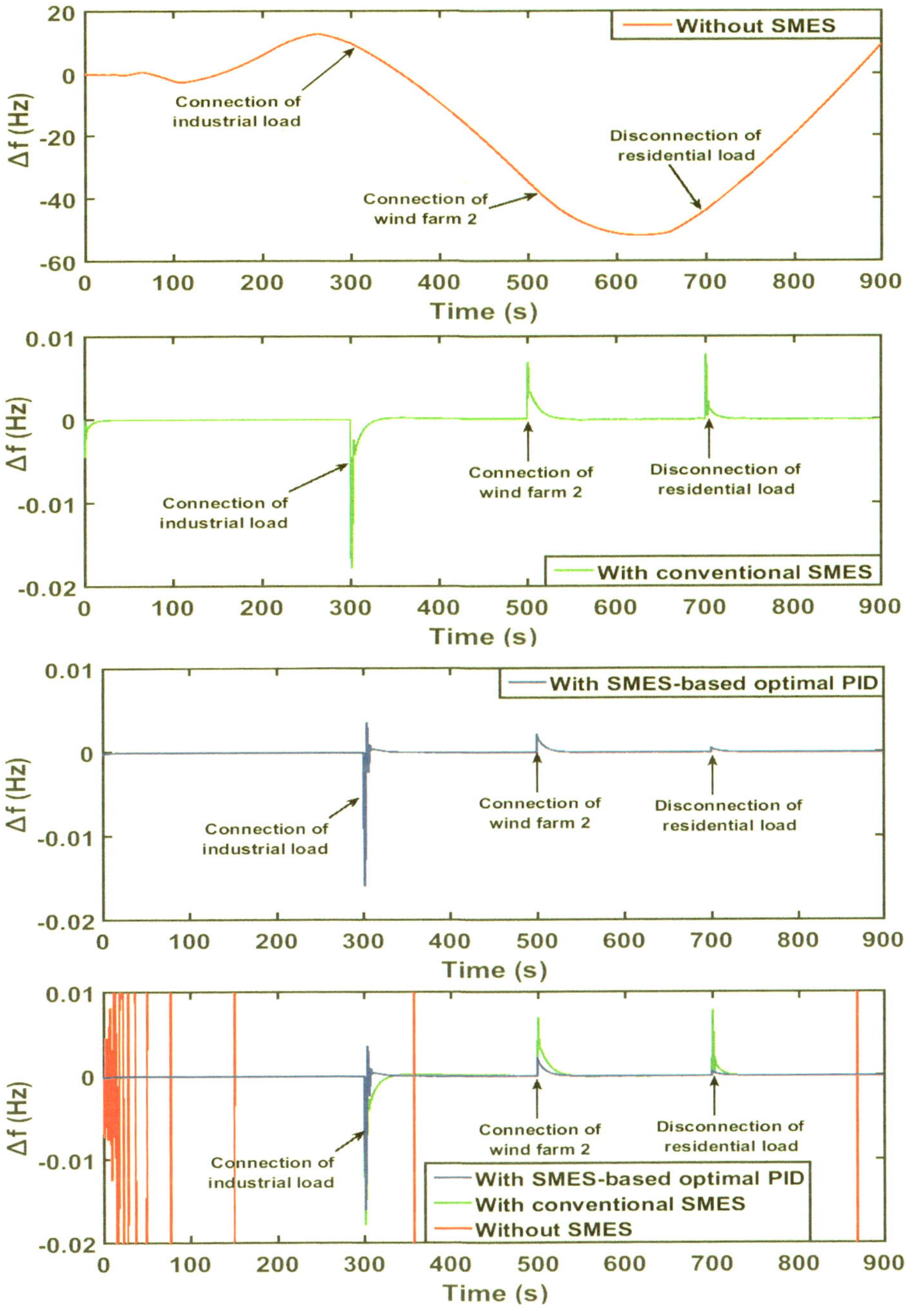

Fig. 2.12 The frequency deviation of the EPS for scenario B3 (i.e., 50% of default system inertia)

Table 2.10 The performance specification of the studied system for scenario B

Scenario B		With SMES-based PID	With SMES	Without SMES
B1	MUS (pu)	0.0081	0.0129	0.0256
	MOS (pu)	0.0020	0.0050	0.0100
	T_S (s)	21	37	45
B2	MUS (pu)	0.0106	0.0170	0.0336
	MOS (pu)	0.0021	0.0056	0.0254
	T_S (s)	24	42	48
B3	MUS (pu)	0.0157	0.0197	–
	MOS (pu)	0.0032	0.0068	–
	T_S (s)	25	46	–

the aim of both frequency regulation (i.e., LFC) and enhancement of the input control signal to SMES. To prove the robustness of the proposed coordination, the EPS was analyzed in the presence of high renewable energy penetration using different scenarios considering the multiple operating conditions of wind and load variations. The results observed by simulations showed that the proposed coordination achieved robust frequency stability in the presence of HWPP and different load power fluctuations against all cases of studied scenarios in terms of peak overshoot, peak undershoot, and settling time. The studied power system in Egypt with the dynamic contribution of conventional SMES as a frequency stabilizer model gave a satisfactory performance but needs a little long time to suppress the frequency deviations compared with the proposed coordination. On the other hand, in case of no SMES controller, the frequency fluctuates more with high deviation. In addition, the frequency oscillates to such an extent that it is not acceptable in case of large disturbances, which happens by decreasing the system inertia, resulting from the power electronic interface-based RESs.

Chapter 3
A Comprehensive Digital Protection Scheme for Low-inertia Microgrids Considering High Penetration of Renewables

3.1 Introduction

In the past, several cascaded blackouts happened in electrical power systems due to frequency instability in case of the imbalance between the electrical load and power supply or N-1 contingency [84]. Nowadays, this problem has increased after the growth of RESs which have several impacts on the performance of the islanded μGs such as the reduction of system inertia. Consequently, there is increase in the voltage and frequency fluctuations [1]. Furthermore, the RESs exchange electrical power to μGs through power electronic inverters, which cause higher power fluctuations than the traditional SGs. Therefore, if the RES penetration becomes larger, the islanded μGs might become insecure as stabilizing of system frequency and voltage is difficult in that situation [2, 85]. Moreover, there will be unbalance between the generation and load due to the variable nature of RESs. These changes lead to the appearance of challenges for the conventional control and protection systems such as nature transient variations in μG. These variations are highly affected by the operation mode of μG whether grid-connected or stand-alone [44]. On the other hand, the selection and coordination of conventional protective relays become more complex due to the frequent bidirectional power flow in connection feeders of μGs to utility grids [86]. Therefore, many studies have developed different control strategies to deal with these challenges. However, in case of large frequency deviations, which might be followed by an additional generation or load events leading to imbalances in active power during the fault periods, the control technique is unable to restore the system frequency. In that situation, the emergency control and protection must be used to return the system frequency to its steady-state condition.

The protection and control are two main sides dealing with these frequency instability problems. They have changed significantly from the bygone decade and will change continuously as a result of the advancement of technology. Therefore, the power system designers have sought to apply digital devices to handle the increase in power system complexity. The digital world has impacted the protection

G. Magdy et al., *Renewable Power Systems Dynamic Security*, Power Systems,
https://doi.org/10.1007/978-3-030-33455-0_3

system from the introduction of microprocessor-based relays in the 1980s to protection relays with communication interfaces in the 1990s [87]. Today, digital relays are featured with high-speed communication, which helps in replacing wires for safety interlocking, control, and also circuit breaker tripping action. Furthermore, there are many applications of digital relays in transmission and generation system protection due to their flexibility, high-performance level, and capability of operating under different temperatures compared to the classical electromechanical relays. On the other side, the discrete-time controllers become more appealing to replace analog controllers in different power systems. Using digital control systems reduces the implementation cost and increases the reliability of the control system [88]. Therefore, many digital devices have been used in electrical power systems such as digital PID controller, digital power system stabilizer (PSS), digital automatic voltage regulator, and digital protection devices, i.e., digital OUFR. There are two approaches for designing digital control systems. The first one is the direct digital design approach, which converts the analog plant to discrete and then defines a digital controller for the discretized plant. The second approach is the digital redesign approach [72, 88], which designs a good analog controller for the analog plant and then carries out the digital redesign for the good designed analog controller. This study focuses on the second approach, which is indirect design approach due to the following: (a) it is more realistic to carry out the design of the continuous-time controller and (b) it is easy to select the sampling time which can be selected after defining the continuous-time closed-loop bandwidth.

According to the most recent researches, frequency regulation issue of an islanded μGs utilizes several types of control strategies, such as conventional controller with different algorithms and optimization techniques [89], and intelligent control, i.e., fuzzy logic control (FLC) [31], MPC [45], and robust control [44, 67]. However, application of discrete-time controllers to power system was reported in several types of research [72, 88, 90, 91]. Shabib and Hori [90] presented a discrete-time model of continuous-time PSS for the transient stability of single machine infinite bus utilizing Tustin's discretization method. However, this model of digital PSS withstands the large disturbances for small sampling intervals, which needs hardware devices with a high cost. Dashti et al. and Nadra et al. [91] applied a new digital redesign approach for the conventional PSS discretization based on the optimal matching of the continuous-time closed loop. The new approach used in [91] is called plant input mapping (PIM) method. However, the optimization problem is solved by sampled data control theory. Moreover, Shabib et al. [72, 88] used the same technique for discretization continuous-time conventional controllers for a single machine power system to guarantee the stability of any sampling rate.

On the other hand, several studies have dealt this problem from the short circuit fault side only such as the optimized time-based coordination of conventional overcurrent relays, which is the earliest protection technique for utility grids including microgrids [92]. This method has a limit in its ability of multi-relay protection because of its high sensitivity to component parameters in high fault levels. Sheng et al. [93] presented a multi-agent method depending on assumptions of high fault current levels. But this method has been developed to island the μG for any fault in the utility grid and also disconnect most of the DGs for faults within the μG. Furthermore, some studies handled the frequency protection problems such as that of Laghariet et al. [94] which applied an intelligent computational technique for load shedding of the power system under faulted conditions. Moreover, Tephiruk et al. [95] discussed the same issue using the rate of change of underfrequency relay to improve the load shedding scheme in μG systems. Further, Vieira et al. [96] proposed the coordination of ROCOF and OUFRs. However, this coordination did not compensate the frequency fluctuations within the allowable frequency limit because the action of the relay is energized once the system frequency becomes out of the allowable limit. Such a problem can be overcome by designing the proposed coordination of frequency stability using digital LFC and digital OUFR. Therefore, this chapter presents the design of digital OUFR coordinated with the digital LFC for a small μG system, which consists of thermal power plant, PV, wind power generation (WPG), and domestic loads. To prove the effectiveness of the proposed coordination for the islanded μG against frequency variations, it has been tested under different scenarios of disturbances such as high penetration level of RESs, reducing system inertia, and load shedding/restoring.

3.2 Problem Description

Since the power generation from RESs is unpredictable and variable, it results in more fluctuations in power flow and frequency in the μG, which significantly affects the power system operation. Therefore, the high penetration of RESs makes the situation in μG worse because of the low inertia and small time constant of the system, consequently creating difficulty in stabilizing system frequency and voltage, causing the weakening of μG stability and resiliency. Moreover, the random changes in load power demand cause a bad response to the point of common coupling voltage, active, and reactive power transfer. Hence, a severe frequency deviation in the presence of high RES power fluctuations has an adverse impact on the control performance parameters and may cause energizing of under/overfrequency relay and disconnect some loads and generation parts of the μG system. Therefore, the stability and protection coordination issues have become a centre of interest, especially for power system researchers.

3.3 System Configuration

3.3.1 Structure of μG

The μG is a small power system, which contains DG units, domestic loads, EESs, and power-conditioning units. The μG is distributed through low-voltage distribution systems and the electric power is mainly generated by DGs such as PV, wind

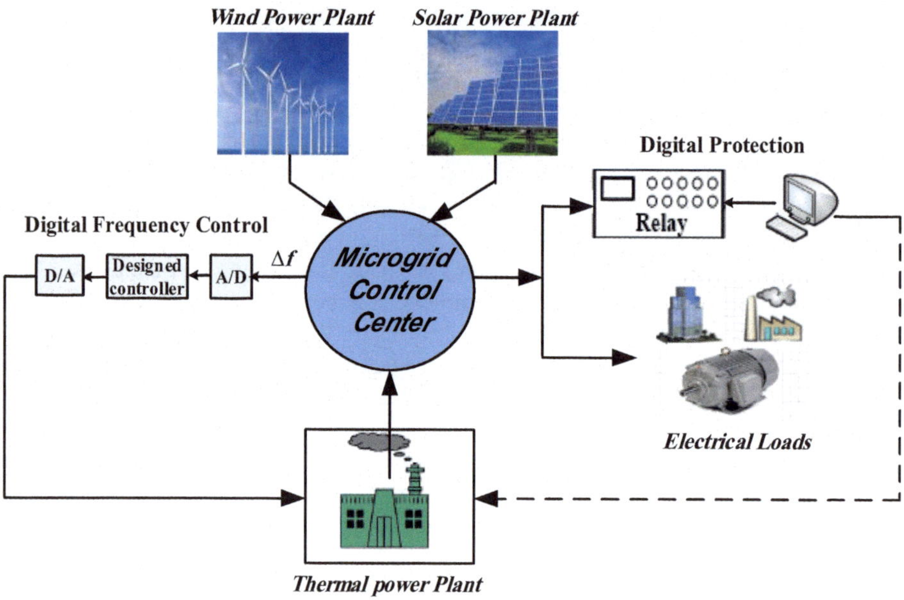

Fig. 3.1 An islanded μG system with digital coordination strategy

turbines (WT), hydro units, and fuel cells. This study focuses on the islanded μG (base of 20 MW), which includes 15 MW of domestic loads, 20 MW of thermal power plant, 6 MW of a wind farm, and 4.5 MW of a solar farm as shown in Fig. 3.1.

In this study, the effects of the physical constraints such as GRC of power plants and GDB are taken into consideration for modeling the actual islanded μG. The GRC limits the generation rate of output power which is given as 0.2 p.u. MW/min for the non-reheat power plant. The V_U and V_L are the maximum and minimum limits that restrict the rate of the valve-gate closing or opening speed [44]. In this study, the power variation of RESs such as the wind power variation (ΔP_{Wind}), the PV solar power variation (ΔP_{PV}), and the load power variation (ΔP_L) are considered as disturbance signals for islanded μG. The dynamic model of the studied μG system is shown in Fig. 3.2. The μG nominal parameter values are shown in Table 3.1.

3.3.2 *Mathematical Model of the Islanded μG*

The different subsystems of the islanded μG in Fig. 3.2 can be described below.

1. Governor model: The speed governor adjusts the turbine gate to return the frequency to its nominal value. The governor is modeled by a first-order function of a unity gain and time constant (T_g):

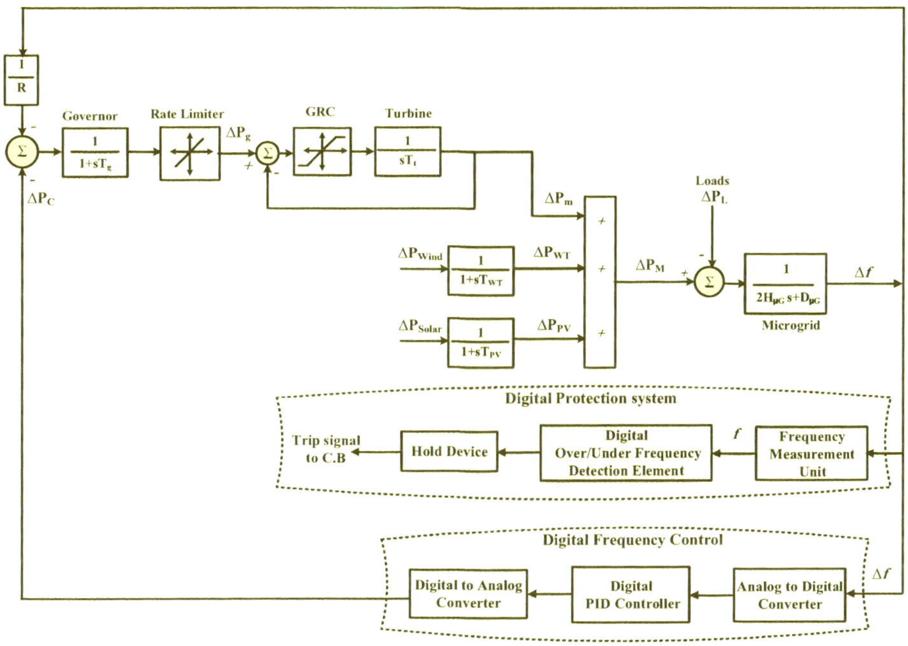

Fig. 3.2 The dynamic model of the islanded μG with the proposed coordination

Table 3.1 Islanded μG parameters [97]

Parameter	Value	Parameter	Value
$D_{\mu G}$	0.015	T_{PV}	1.8
$H_{\mu G}$	0.083	R	2.4
T_g	0.1	V_U	0.3
T_t	0.4	V_L	−0.3
T_{WT}	1.5	GRC	20%

$$\mathrm{TF}_g = \frac{1}{1 + sT_g} \tag{3.1}$$

where the governor with speed droop ($1/R$) represents the primary frequency control of governor action.

2. Turbine model: The turbine model considered in this study is a steam turbine, which represents a first-order transfer function as this relation:

$$\mathrm{TF}_t = \frac{1}{1 + sT_t} \tag{3.2}$$

In order to get an accurate perception of the actual μG, this study considers the nonlinear model, that is, considering various effects of physical system dynamics, including the GRC and maximum/minimum turbine limits [1].

3. Rotating mass and load: This block represents the machine's mechanical dynamic loop, which is modeled by a first-order transfer function of system inertia constant ($H_{\mu G}$) and damping coefficient ($D_{\mu G}$) as seen in the following equation:

$$\mathrm{TF}_r = \frac{1}{2H_{\mu G}s + D_{\mu G}} \tag{3.3}$$

4. Wind power system: This study uses an aggregated model to form the wind farm as a large-capacity induction generator. The generator is modeled by a given first-order lag transfer function with a unity gain and time constant (T_{WT}), neglecting all nonlinearities, as given below:

$$\mathrm{TF}_{\mathrm{WTG}} = \frac{\Delta P_{\mathrm{WT}}}{\Delta P_{\mathrm{Wind}}} = \frac{1}{1 + sT_{\mathrm{WT}}} \tag{3.4}$$

5. Solar PV system: The model of solar power is presented as a disturbance source to the islanded μG. Therefore, it is modeled by a simple linear first-order lag of a unity gain and time constant (T_{PV}):

$$\mathrm{TF}_{\mathrm{PV}} = \frac{\Delta P_{\mathrm{PV}}}{\Delta P_{\mathrm{Solar}}} = \frac{1}{1 + sT_{\mathrm{PV}}} \tag{3.5}$$

Hence, the overall generator–load dynamic relation between the net power and the frequency deviation (Δf) can be obtained as

$$\dot{\Delta f} = \frac{1}{2H_{\mu G}}(\Delta P_m + \Delta P_{\mathrm{WT}} + \Delta P_{\mathrm{PV}} - \Delta P_{\mathrm{L}}) - \frac{D_{\mu G}}{2H_{\mu G}} * \Delta f \tag{3.6}$$

where

$$\Delta\dot{P}_{\mathrm{g}} = -\frac{1}{T_{\mathrm{g}}}(\Delta P_{\mathrm{g}}) - \frac{1}{R.T_{\mathrm{g}}} * \Delta f + \frac{1}{T_{\mathrm{g}}}(\Delta P_{\mathrm{C}}) \tag{3.7}$$

$$\Delta\dot{P}_{\mathrm{m}} = -\frac{1}{T_t}(\Delta P_{\mathrm{m}}) + \frac{1}{T_t}(\Delta P_{\mathrm{g}}) \tag{3.8}$$

$$\Delta\dot{P}_{\mathrm{PV}} = \frac{1}{T_{\mathrm{PV}}}(\Delta P_{\mathrm{Solar}}) - \frac{1}{T_{\mathrm{PV}}}(\Delta P_{\mathrm{PV}}) \tag{3.9}$$

$$\Delta\dot{P}_{\mathrm{WT}} = \frac{1}{T_{\mathrm{WT}}}(\Delta P_{\mathrm{Wind}}) - \frac{1}{T_{\mathrm{WT}}}(\Delta P_{\mathrm{WT}}) \tag{3.10}$$

where ΔP_{wind}, $\Delta P_{\mathrm{solar}}$, and ΔP_{L} are wind power variation, solar power variation, and load power variation, respectively. These variations are considered as the μG disturbance signals. The damping ($D_{\mu G}$) and the inertia ($H_{\mu G}$) are the uncertainty parameters. Δf is the frequency deviation, ΔP_{m} is the thermal power deviation, and

ΔP_g is the governor power deviation. Therefore, the dynamic equations of the studied hybrid power system can be derived and written in the state variable form as follows:

$$\dot{X} = \mathrm{AX} + \mathrm{BU} + \mathrm{EW} \tag{3.11}$$

$$Y = \mathrm{CX} + \mathrm{DU} + \mathrm{FW} \tag{3.12}$$

The complete state-space model of the presented μG considering high RES penetration level can be obtained through the state variables and definitions from Eqs. (3.6)–(3.10). The linearized state-space model of the μG from Fig. 3.2 is as in Eqs. (3.13) and (3.14):

$$\dot{X} = \begin{bmatrix} -\frac{D_{\mu G}}{2H_{\mu G}} & 0 & \frac{1}{2H_{\mu G}} & \frac{1}{2H_{\mu G}} & \frac{1}{2H_{\mu G}} \\ -\frac{1}{\mathrm{RT_g}} & -\frac{1}{T_g} & 0 & 0 & 0 \\ 0 & \frac{1}{T_t} & -\frac{1}{T_t} & 0 & 0 \\ 0 & 0 & 0 & -\frac{1}{T_{WT}} & 0 \\ 0 & 0 & 0 & 0 & -\frac{1}{T_{PV}} \end{bmatrix} * \begin{bmatrix} \Delta f \\ \Delta P_g \\ \Delta p_m \\ \Delta p_{WT} \\ \Delta P_{PV} \end{bmatrix} + \begin{bmatrix} 0 \\ \frac{1}{T_g} \\ 0 \\ 0 \\ 0 \end{bmatrix} * [\Delta P_C] + \begin{bmatrix} 0 & 0 & -\frac{1}{2H_{\mu G}} \\ 0 & 0 & 0 \\ 0 & 0 & 0 \\ \frac{1}{T_{WT}} & 0 & 0 \\ 0 & \frac{1}{T_{PV}} & 0 \end{bmatrix} * \begin{bmatrix} \Delta P_{Wind} \\ \Delta P_{Solar} \\ \Delta P_L \end{bmatrix} \tag{3.13}$$

$$Y = [1\ 0\ 0\ 0\ 0\ 0] * \begin{bmatrix} \Delta f \\ \Delta P_g \\ \Delta p_m \\ \Delta p_{WT} \\ \Delta P_{PV} \end{bmatrix} + [0][\Delta P_C] + [000] \begin{bmatrix} \Delta P_{Wind} \\ \Delta P_{Solar} \\ \Delta P_L \end{bmatrix} \tag{3.14}$$

Table 3.2 Frequency operation and control/protection actions [97]

Frequency deviation	Condition	Action
Δf_1 (0.3 Hz)	No contingency or load event	Primary control
Δf_2 (1 Hz)	Generation/load event	Secondary control (i.e., LFC)
$\Delta f_3 >$ (2 Hz)	Large separation event	Protection operation

3.4 Proposed Digital Coordination Strategy

3.4.1 Control Scheme

1. Frequency control of an islanded μG:
   The power system frequency may have high variations if there is no longer a balance between the generation and load demand. The normal frequency deviations can affect the power system's efficiency and reliability, while large deviations can destroy the equipment, overload transmission lines, and cause interference with the system protection. Therefore, the frequency control is divided into three main operations based on the size of the frequency deviations. The frequency deviation ranges and their control actions are shown in Table 3.2. However, the normal frequency deviations up to Δf_1 and the power requirement are balanced by attenuating these deviations by the governor natural autonomous, which is named primary control. If the frequency deviations are more than Δf_1 up to Δf_2, the secondary frequency control (i.e., LFC) must recover the system frequency to its steady-state condition within the limits of standard time deviations. However, in case of large frequency deviations such as Δf_3 and over, which lead to imbalances in active power during the fault periods, the LFC cannot maintain the system frequency. In that situation, the protection devices (i.e., frequency relays) may be activated and trip generators. This action will interrupt power system supply. Hence, there must be accurate coordination of LFC or emergency control and protection scheme.
2. Digital Frequency Control system:
   In this study, a discrete-time PID controller presents as LFC for regulation of the islanded μG frequency. The discretization process of PID controller is implemented using the mapping technique, which has three classical discretization approaches for converting a model from S-domain to Z-domain. The three approaches are (a) forward difference method, (b) backward difference method, and (c) bilinear transformation method or in practice called Tustin's method [90]. This study focuses on the last approach due to its advantages compared to other approaches such as easy implementation and convergence to analog one. Moreover, in this technique, the left-hand side of the S-plane is mapped within the unit circle in the Z-plane. The discrete-time PID controller approximation is obtained for the transfer function of the continuous-time PID controller simply by replacing S-domain to Z-domain, according to this relation:

Table 3.3 PID controller's parameters for the μG

PID parameters		
K_p	K_I	K_d
9.68204	0.806941	18.73075

$$s = \frac{1}{T} \ln z = \frac{z - 1}{\frac{T}{2} * (z + 1)} \tag{3.15}$$

where T is the sampling interval of the discrete-time system, which is selected as ($T = 0.01$ s) for this study. The design parameters of the digital PID controller for regulating the frequency of islanded μG are given in Table 3.3.

3.4.2 Protection Scheme

1. Modeling of digital frequency relay:
   The frequency relay is a member of the protection device group. It is used to protect the power system from a blackout in case of load loss, generation loss, or N-1 emergency. Furthermore, it is used in the μG network to detect the islanding operation, which occurs in the case of DGs because of losing of mains [98]. Moreover, the main threat occurs when a DG is reconnected to the rest of the system without synchronizing operation at first. In the past, DGs are directly disconnected from the system due to over- or underfrequency problems. Recently, the continuous operation of DGs to supply domestic loads in islanded condition becomes necessary. Therefore, the use of digital relays has spread and become more widely used in μGs as the digital relays can change their settings according to the abnormality conditions. Furthermore, recently, there are many applications of digital relays in transmission and generation system protection due to their advantages such as flexibility, high performance level, and capability of operating under different temperatures compared to the classical electromechanical relays [99]. The digital relay is a basic component of the digital protection system as shown in Fig. 3.3. The digital frequency protection system comprises four basic components: instrument transformers (e.g., potential transformer) that measure the values of current and voltage, frequency measurement unit (FMU) that measures the value of system frequency, frequency detection element (FDE) that processes the received data utilizing the objective function algorithms such as over/underfrequency protection, and data conversion system [100]. Therefore, at the point when an irregular condition is recognized, the relay detects the problem and sends a trip signal to the circuit breaker. As shown in Fig. 3.2, the change of mechanical power (i.e., $\Delta P_M = \Delta P_m + \Delta P_{WT} + \Delta P_{PV}$) will balance the change of the power demand (ΔP_L) at the steady-state operation of the islanded μG as per swing equation as shown below:

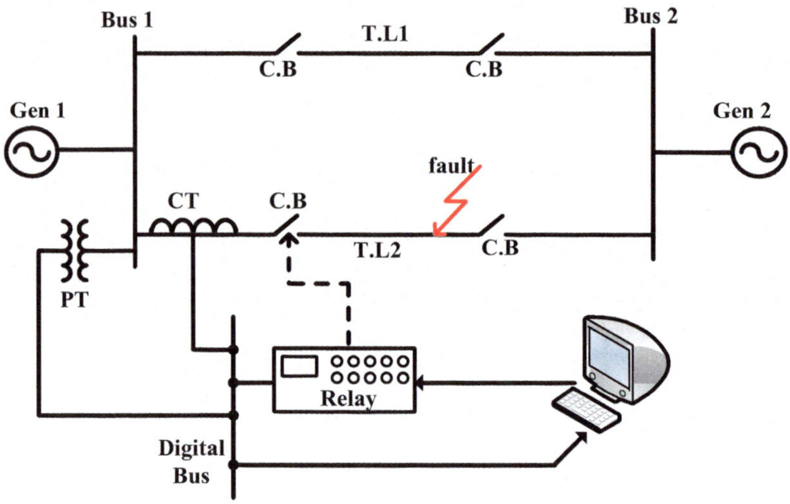

Fig. 3.3 A simplified digital protection system [97]

$$\frac{2H}{\omega_o}\frac{\mathrm{d}\omega}{\mathrm{d}t} = \Delta P_{\mathrm{M}} - \Delta P_{\mathrm{L}} \tag{3.16}$$

The rate of angular speed change can be inferred as given in Eq. (3.17). Therefore, it is easy to calculate the RoCoF of the power system to estimate the frequency relay setting according to Eq. (3.18). Moreover, the digital OUFR should be adjusted with the time delay setting (i.e., integrator action); thus, the system frequency deviation continues during a pre-characterized time interim for tripping the circuit. Hence, the setting of the delay time can be calculated as in Eq. (3.19):

$$\Delta\omega = \frac{\omega_o\left(\Delta P_{\mathrm{M}} - \Delta P_{\mathrm{L}}\right)}{2H}t \tag{3.17}$$

$$\Delta f = \frac{f_o\left(\Delta P_{\mathrm{M}} - \Delta P_{\mathrm{L}}\right)}{2H}t \tag{3.18}$$

$$t = \frac{2H\Delta f}{f_o(\Delta P_{\mathrm{M}} - \Delta P_{\mathrm{L}})} + K \tag{3.19}$$

where K is the threshold time for the integrator, where it determines the relative operating time of the OUFR. Therefore, the settings of the relay (i.e., $f_{\max}$, $f_{\min}$, and K) are selected according to country standards for the power system operation [100]. Therefore, according to the European grid code for the power system operation [101], the setting of the proposed OUFR is indicated in Table 3.4.

2. Principal operation of digital frequency relay:

Table 3.4 Frequency relay settings

Frequency relay settings	Limit	Threshold time (K)
Overfrequency relay	$F_{max} = 51$ Hz	5 s
Underfrequency relay	$F_{min} = 49$ Hz	

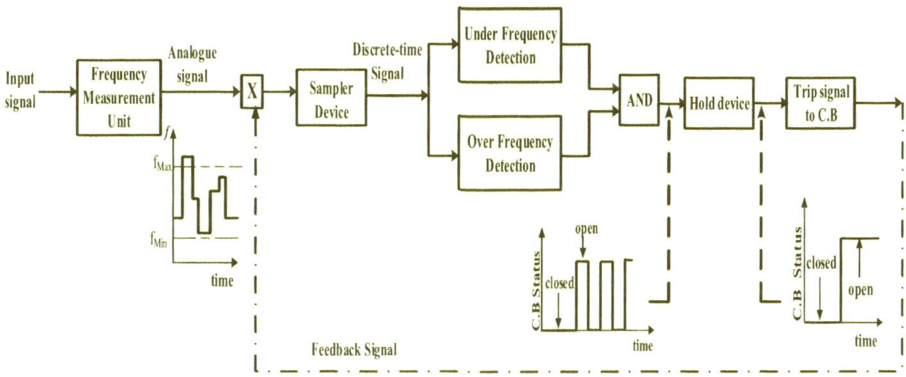

Fig. 3.4 The logic diagram of the digital frequency protection system [102]

The operation procedures of the digital frequency protection system are (1) measuring the system frequency of a voltage signal through the FMU, (2) processing the received data (i.e., the system frequency) and taking the necessary action using the algorithm of over/underfrequency protection through the FDE, and (3) sending a trip signal to the circuit breaker if the over/underfrequency conditions occur (i.e., the system frequency overtook the allowable limits) [100]. Therefore, the logic diagram for the implementation of the digital frequency protection system is shown in Fig. 3.4. Moreover, the sequence of the algorithm of the over/underfrequency protection is introduced in Fig. 3.5, where the operation of the OUFR depends on measuring the system frequency and is compared it with the frequency limits of the OUFR ($f_{max} < f < f_{min}$). To energize the relay and send a trip signal to the circuit breaker, two conditions must be met together, namely (1) exceeding the system frequency and the permissible limits of the over- and underfrequency setting and (2) integrator output magnitude being larger than the integrator set time ($K = 5$ s). On the other hand, the roles of the frequency control loops, the primary frequency control (i.e., governor action), and secondary frequency control (i.e., LFC) emerge when the system frequency becomes out of the allowable frequency limits as well as the integrator output magnitude doesn't exceed the integrator set time (K). Hence, the operation of digital OUFR is concluded in the flowchart in Fig. 3.6.

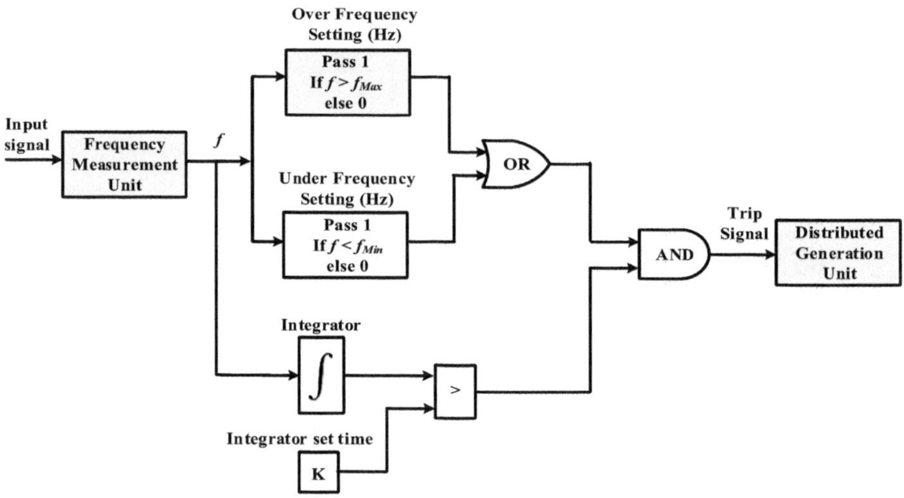

Fig. 3.5 The diagram of the over/underfrequency protection [102]

3.5 Results and Discussion

The proposed digital coordination of LFC and OUFR protection is tested on the single-line diagram of the islanded μG in Fig. 3.7 under the nature variety RESs, random load variation, and system parameter variations, which are known as the important characteristics of an actual μG. The simulation results and analysis of the islanded μG frequency during multiple changes in wind power generation (WPG), solar power, domestic loads (i.e., disturbances), system inertia, and parameters (i.e., uncertainties) are carried out using MATLAB/Simulink. The wind power with 5 MW (i.e., 25% from the system base) is integrated to the islanded μG at 500 s, while the PV solar power with 3 MW (i.e., 15% from the system base) is connected from the initial time. The islanded μG is tested in the presence of high-fluctuated wind power and low-fluctuated solar power as shown in Fig. 3.8 for a simulation time of 15 min. To investigate the effectiveness of the proposed digital coordination on the islanded μG frequency response, five several scenarios are applied to the μG as follows:

Scenario A: In this scenario, the effectiveness of the proposed digital coordination for the islanded μG is evaluated by implementing the random domestic load variations as shown in Fig. 3.9a, in addition to connecting the high-fluctuated wind power and low-fluctuated solar power. In this case, the variation of the system frequency is the second type of frequency deviation (Δf_2) and within the limits of the digital relay. The digital frequency relay does not trip as seen in Fig. 3.9b because the integrator output value does not exceed the set value. Therefore, the digital LFC succeeded to readjust the frequency to its normal value as shown in Fig. 3.9c. This case proves the effectiveness of the digital LFC as it can adjust the frequency to its normal value in all five stages of this scenario without the need for a protection action.

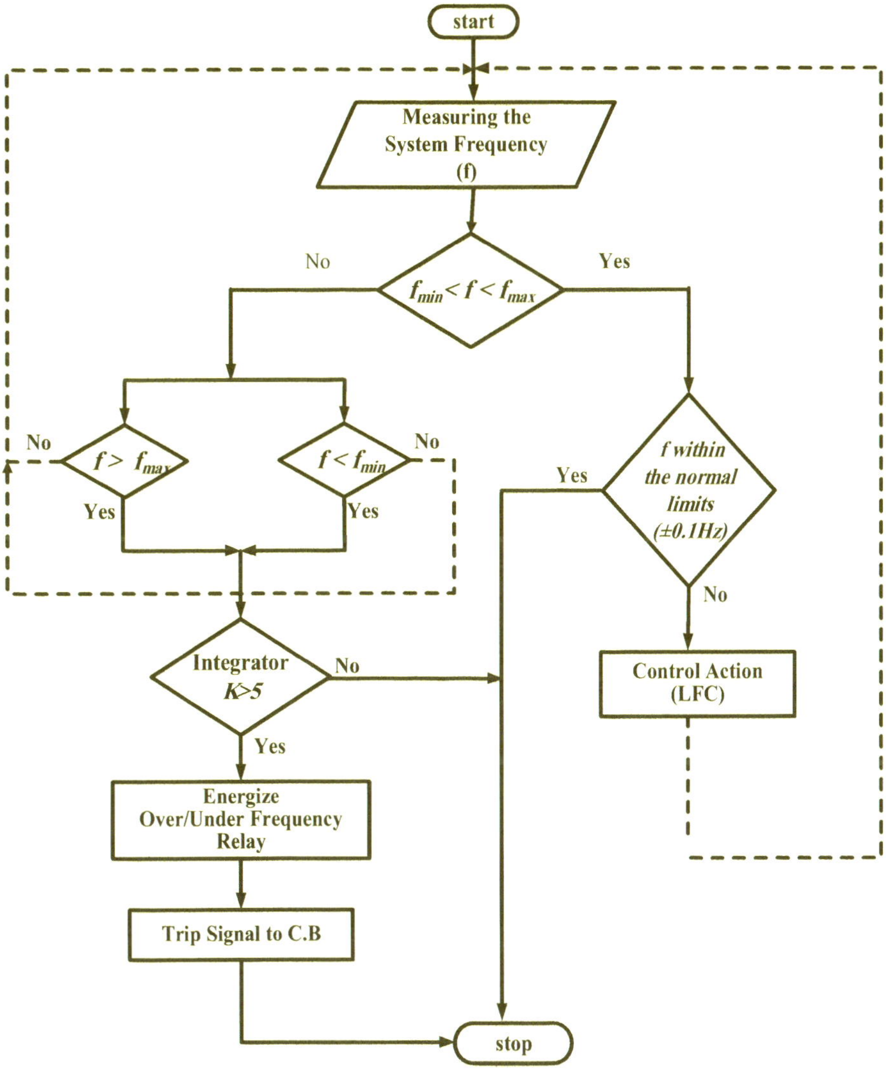

Fig. 3.6 Flowchart of the proposed coordination

Scenario B: In this scenario, the μG system is subjected to the power change under different load disturbance profiles as shown in Fig. 3.10a besides the power fluctuations from wind and PV sources. The digital LFC can handle the frequency deviations and succeed to restore the μG frequency to its normal value during the first load change at 300 s and the instant of wind farm connection at 500 s as seen in Fig. 3.10c. Hence, there is no need for relay action. On the other hand, the digital LFC is unable to control the frequency when the heavy load of 40% is applied at 700 s as the system frequency fluctuates beyond the digital relay setting limits.

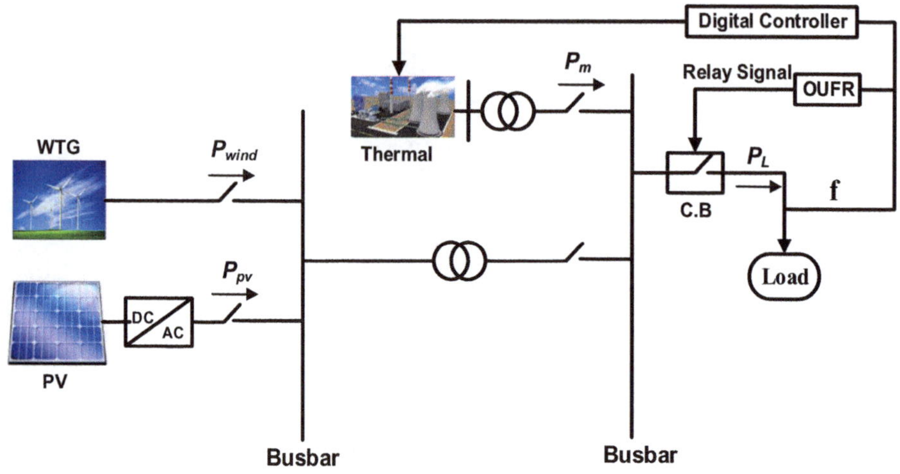

Fig. 3.7 Single-line diagram of the μG case study

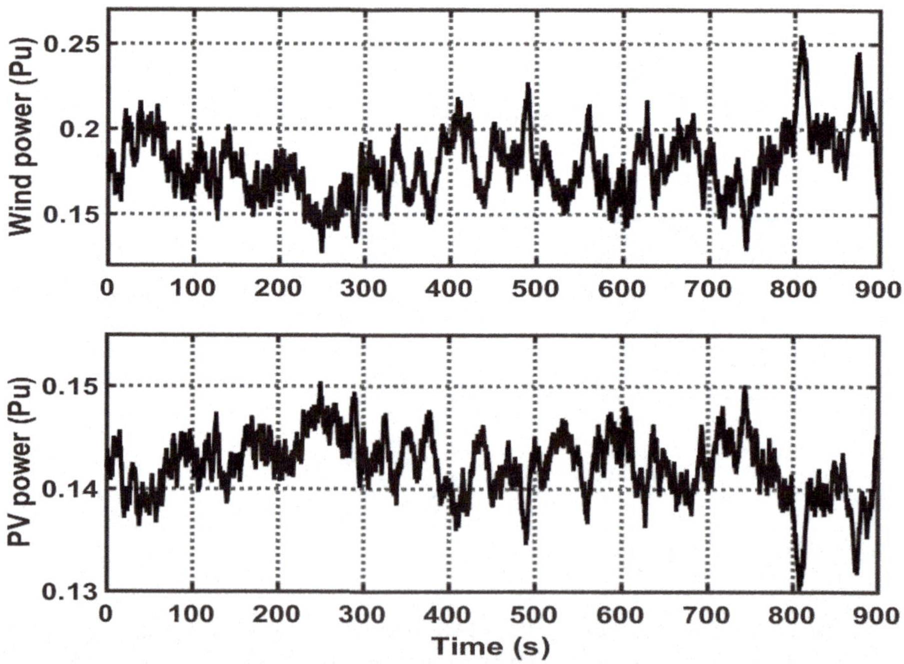

Fig. 3.8 Power variation pattern of wind and solar generations

Furthermore, the integrator output exceeds the integrator set time *K*. Therefore, the digital relay is energized and sends a trip signal to the generator circuit breaker in this case as shown in Fig. 3.10b.

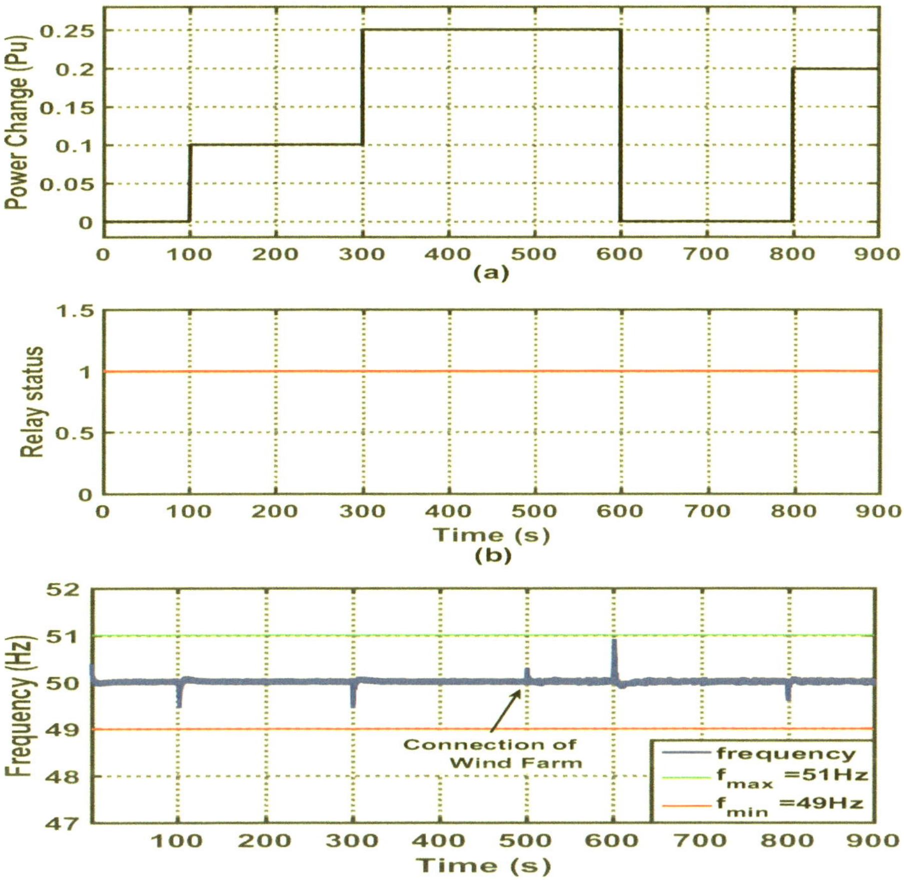

Fig. 3.9 Scenario A: (**a**) Load disturbance, (**b**) relay status, (**c**) frequency response of μG

Scenario C: In this case, the islanded μG behavior is tested for implementation of the same profile of load disturbance as the previous scenario as depicted in Fig. 3.11a. However, the penetration of wind power increased to 35% (7 MW) from the system base. It is integrated into the μG at 500 s. The digital LFC has the ability to control and restore the frequency to its steady-state value at the first load disturbance at 300 s, while it cannot withstand the change of system frequency caused by high wind penetration at 500 s as noted in Fig. 3.11c. Hence, the digital OUFR sends a trip signal to the generator circuit breaker at that time as shown in Fig. 3.11b, whereas the integrator output exceeds the threshold value of 5 s. Hence, the effectiveness of the proposed coordination is approved.

Scenario D: To evaluate the behavior of the μG under the system uncertainties, the μG is tested under the situation of half of the system inertia (50% of the default value) with multiple operating conditions of wind power and PV power, and load disturbance profile as shown in Fig. 3.12a. The effect of half system inertia through

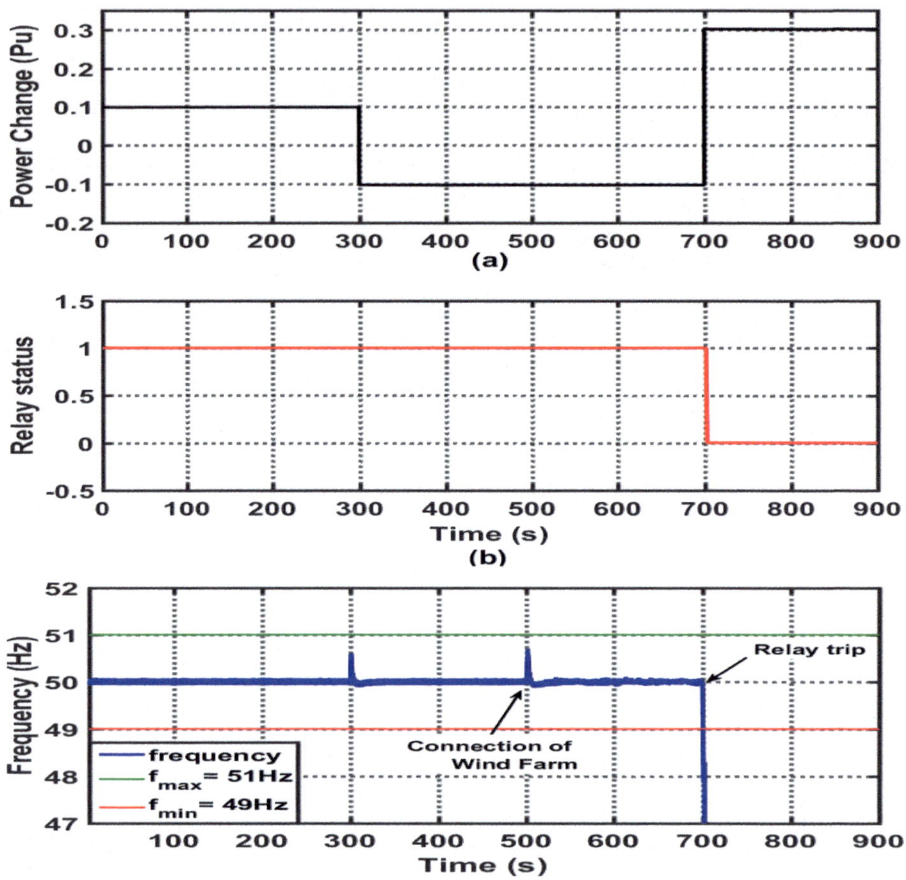

Fig. 3.10 Scenario B: (**a**) Load disturbance, (**b**) relay status, (**c**) frequency response of μG

the proposed digital coordination of LFC and OUFR is investigated. The frequency fluctuations increased in this case with high deviation as depicted in Fig. 3.12c. Although the frequency deviations are not within the allowable frequency limits at 500, 600, and 800 s, respectively, the digital LFC can readjust the frequency to its normal value. This happens because the value of the integrator output does not exceed the threshold value. Hence, the digital LFC action returned the frequency to the nominal value without interference of the digital protection side as seen in Fig. 3.12b.

Scenario E: The effectiveness of the proposed digital coordination is approved in this extreme scenario. However, the μG system inertia behavior is tested by implementing the same profile of load disturbance as the previous scenario as depicted in Fig. 3.13a.

Moreover, the μG system inertia is decreased to 40% of the default system inertia with the same conditions of wind power and PV solar power variations. The digital

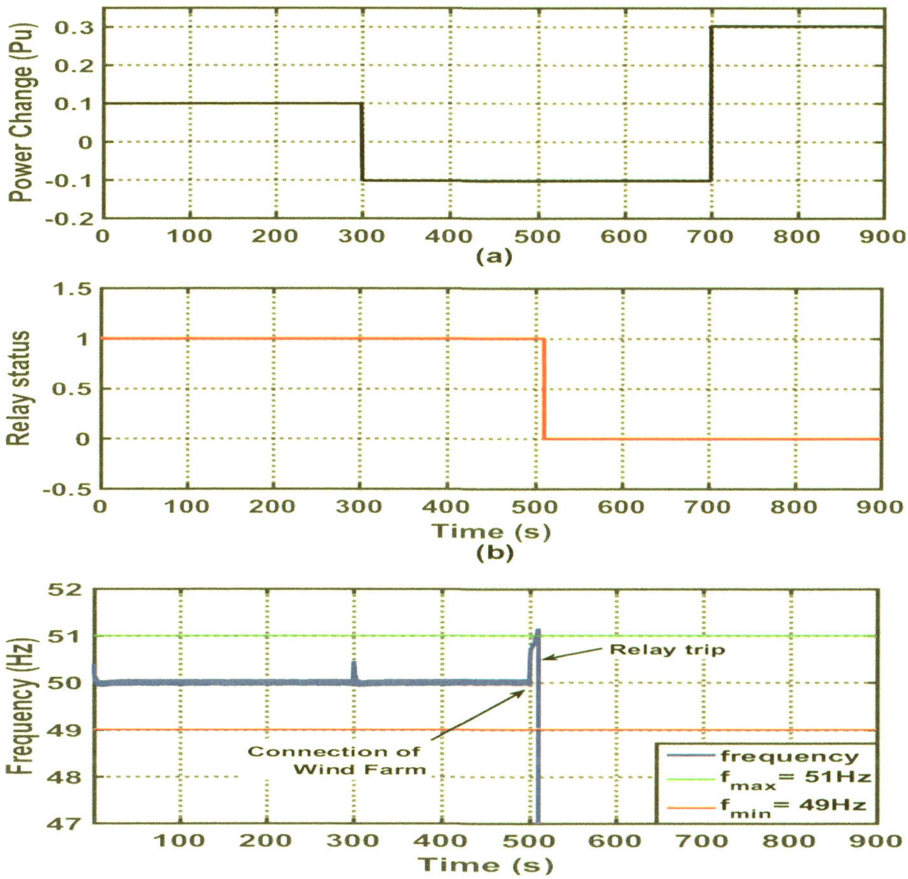

Fig. 3.11 Scenario C: (**a**) Load disturbance, (**b**) relay status, (**c**) frequency response of μG

LFC can withstand the fluctuations and return the frequency to its nominal value at 100, 300, and 500 s, respectively, as shown in Fig. 3.13c, of the μG frequency response. However, it failed to restore the system frequency in the fourth stage at 600 s with more load shedding since the two conditions for energizing the OUFR were achieved: the system frequency was beyond the allowable frequency control limits and the integrator output reached quickly to the threshold value (5 s). Hence, the digital frequency relay trips quickly in this scenario to maintain the equipment from damage as shown in Fig. 3.13b.

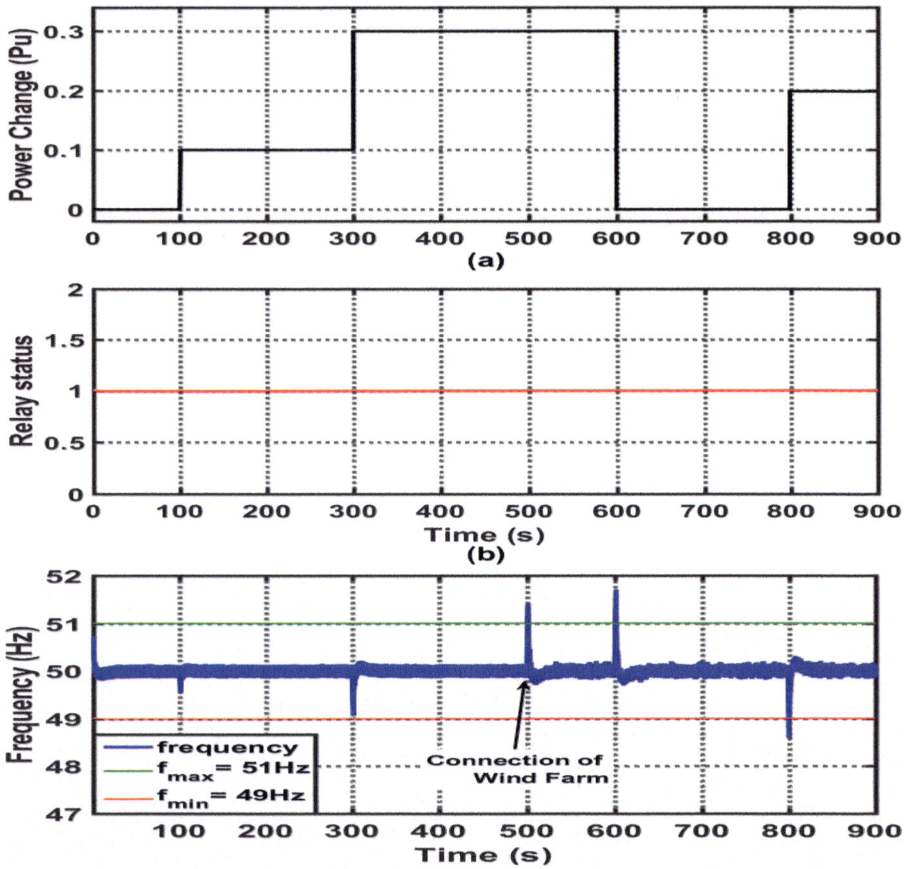

Fig. 3.12 Scenario D: (**a**) Load disturbance, (**b**) relay status, (**c**) frequency response of μG

3.6 Conclusion

This chapter proposed a digital coordination strategy of LFC-based mapping technique and OUFR protection for an islanded μG system security considering high penetration of RESs. This coordination strategy is proposed for supporting the frequency stability and protecting the isolated μG against high-frequency deviations, which increased recently due to the high penetration of RESs, random load variations, and system uncertainty. These changes threaten the μG system security and can lead to cascading outages. The simulations results proved that the proposed digital coordination of LFC and frequency relay has achieved an effective performance for maintaining the system frequency at nominal value. However, the digital LFC succeeded to readjust the frequency deviations to its allowable limits under different conditions of transients, load disturbances, and RES penetration. However, in some cases of large disturbances and high RES penetration, the digital LFC cannot

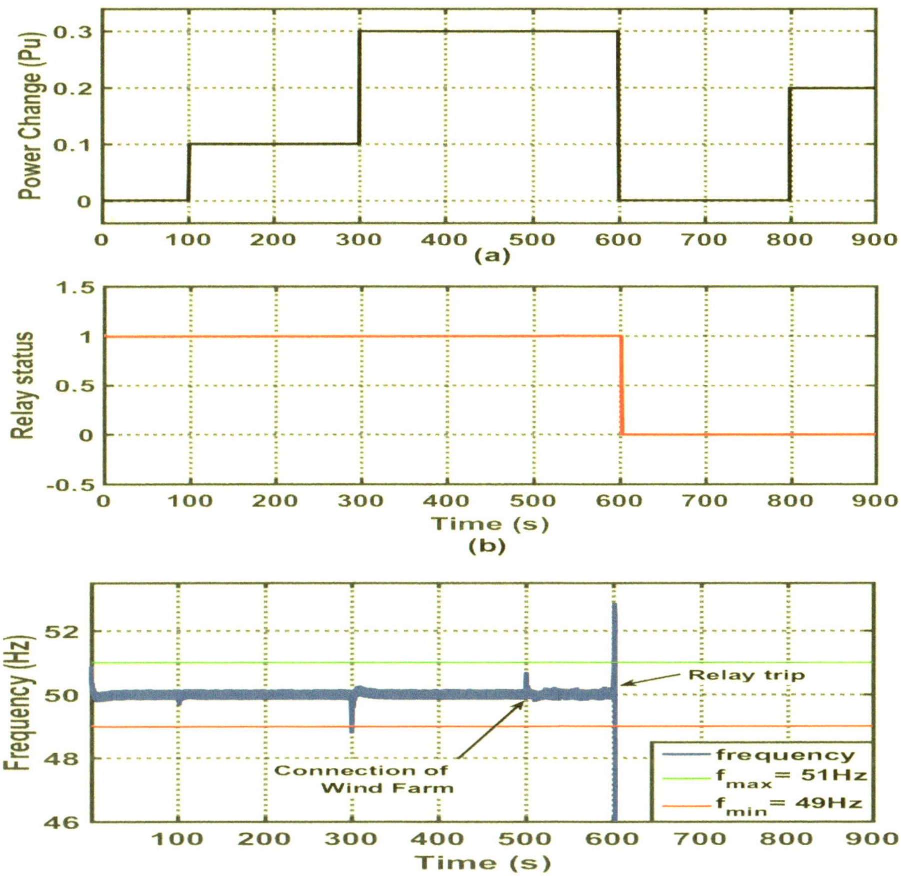

Fig. 3.13 Scenario E: (**a**) Load disturbance, (**b**) relay status, (**c**) frequency response of μG

maintain the frequency stability as the frequency fluctuates beyond the normal limits. In that case, the digital frequency relay will trip the generation unit. Furthermore, the results confirmed that the digital OUFR has superiority in terms of accuracy, sensitivity, and wide range controlling.

Chapter 4
Dynamic Security Assessment of Low-inertia Microgrids Based on the Concept of Virtual Inertia Control

4.1 Introduction

Owing to the recent developments in electricity networks, the use of new and renewable sources of energy such as wind, solar, and geothermal has turned out to be inescapable. However, the intermittent nature of the RESs causes many control problems such as frequency/voltage instability problem, which may be limiting their high penetration [1]. Furthermore, the RESs reciprocate electrical power to the islanded μGs through power electronic devices (e.g., inverters). These inverters do not have any rotating mass, thus lowering the overall system inertia. Consequently, the inverter-based RESs will cause high frequency/voltage fluctuations compared to the conventional SGs [44]. Therefore, if the penetration level in the RESs becomes greater, the islanded μG may become unsafe where the stability of the system frequency/voltage is difficult in this case.

To beat the issue posed by the inverter-based RESs, one of the cutting-edge arrangements is to emulate the behavior of an SG in power systems for enhancing the system inertia, μG stability, and flexibility. It is known as a VSG that imitates the activity of the prime movers [45, 85]. The virtual inertia control is a particular case of VSG execution, where the activity of the prime mover is imitated to support frequency stability [2]. Moreover, it can provide a basis for maintaining the share of DGs in the μG without sacrificing μG stability and flexibility. The virtual inertia control strategy is based on the RoCoF where it calculates the deviation of system frequency to add additional active power to the set point. Thus, this concept simulates the inertia property, which contributes to the total inertia of the μG, enhancing transient frequency stability [103]. On the other side, the protection issue of the μG is considered as one of the most important issues that help to achieve the μG dynamic security. The μG's dynamic security deals with disturbances that result from low system inertia such as large frequency and voltage fluctuations, loss of generation source, forced load shedding, and short circuit faults [3].

G. Magdy et al., *Renewable Power Systems Dynamic Security*, Power Systems,
https://doi.org/10.1007/978-3-030-33455-0_4

Various control techniques have been executed based on the virtual inertia control for frequency stability improvement of the μGs [44–46, 103–107]. In [44], a robust virtual inertia control-based H_infinite technique is proposed for frequency stability analysis of the islanded μG considering high level of the RESs. In [45], the MPC is proposed for the implementation of the virtual inertia control to face the high penetration level of RESs in an islanded μG. The virtual inertia control based on a derivative controller is proposed to enhance the frequency stability of the interconnected power system in [103]. The FLC is proposed for implementation of the virtual inertia control to support frequency stability of the islanded μGs in [104] and [105]. In spite of the control techniques Kerdphol et al. [44], Mentesidi et al. [104], Hu et al. [105] gave a decent dynamic response; they suffer from many drawbacks such as its dependency on the designer's experience and need for long computational time. Although the strategy of the predictive control in [45] has the benefits of quick response and straightforward structure, it needs more time for the calculation of each sampling time. In [46], the virtual inertia control based on frequency response estimation method is applied to improve the system stability with HWPP. On the other hand, the PID or PI controller is considered as one of the most common types of studying the frequency control issue because it has many advantages such as simplicity and low cost [106]. Also, it gives a reliable performance regardless of the disturbances and system parameter variation (i.e., system uncertainties) [72]. In [107], the PI controller is based on virtual inertia control for wind power to enhance the frequency stability of an islanded μG.

On the other side, as a result of technological advances, the protection systems have been changing continuously and periodically since the last decade and to this day. Thence, the designers of today's power systems seek to implement digital devices to cope with the increasing complexity of RPSs, thus improving cost and usability [87]. Nowadays, digital protection devices (e.g., digital relay) have a feature of high-speed communication and hence they aim to replace the wires for interlocking safety, as well as help the control and the tripping action of circuit breakers. In addition, the digital protection devices are used in many applications of power system operation (i.e., generation, transmission, and distribution systems) because they have many merits compared to conventional electromechanical protection devices such as flexibility, reliability, and ability to operate under different temperatures [87]. Therefore, there are many studies that have addressed the protection issue from the perspective of the short circuit fault such as [92] and [93]. In [92], coordination of overcurrent relays based on the optimal time has been presented for utility networks, including μGs. Notwithstanding, this strategy has a high sensitivity to the parameters of the components during high-level faults; thus its ability is limited to protect multiple sequences. An adaptive multi-agent method that depends on presumptions of high levels of fault current is presented in [93]. On the other side, many studies have addressed the protection issue from the perspective of frequency protection such as [94–96] and [108]. In [94], computational intelligence techniques have been applied in the power system for the load shedding. Also, the same strategy that is based on the RoCoF for the underfrequency relay has been applied for protection of the μG in [95]. A comparison between the RoCoF and surge protection relays has been presented for the application of DGs [108]. In any case, the aforementioned studies confronted a hard assignment in relay coordination as their outline may not distinguish the islanded conditions in the requested time. In [96], a coordinated strategy between RoCoF and OUFRs is proposed for the protection of the distributed generation system utilizing the application region. However, the coordination strategy in [96] could not damp the frequency deviation and return it to the allowable limit because of energizing the frequency relay once the system frequency becomes beyond the acceptable limit. Such an issue can be overwhelmed by proposing effective coordination of frequency stability and protection as addressed in this chapter.

Based on the above analysis, some gaps still need to be filled in the μG dynamic security issue. The μG dynamic security is not achieved yet as the protection issue of the μG has not been considered besides the frequency control issues. Hence, there are two sides for preservation of the μG dynamic security (i.e., system synchronism), which are frequency stability and protection of the electrical systems. Therefore, the main contribution of this study includes the following aspects:

1. Propose a new concept of a μG frequency control incorporation of a virtual inertia control based on the optimal PI controller besides the primary frequency control (i.e., governor action) and secondary frequency control (i.e., LFC) to enhance the frequency stability of the islanded μG considering high penetration level of the RESs.
2. The proposed virtual inertia control based on the optimal PI controller is coordinated with the digital OUFR for enhancement of the frequency stability and maintaining of the μG dynamic security due to high-level RES penetration.
3. The additional controller of the virtual inertia control loop is based on the optimal PI controller, which is optimally designed by using the PSO algorithm to minimize the frequency deviations of an islanded μG, thus enhancing μG stability and resilience.
4. The uncertainties of renewable power generators and load are taken into consideration in the virtual inertia control design procedure. Thus, the proposed coordination technique will guarantee an evasion of system instability and collapse.

4.2 Dynamic Modeling of Islanded μG

4.2.1 Structure of Islanded μG

The μG is a small power system that consists of two types of DGs and domestic loads (i.e., residential and industrial loads). The first type of DGs is RESs such as wind, solar, and geothermal, which have uncertain outputs (i.e., non-dispatchable DGs). The second type depends on controllable DGs such as small thermal power plant. This study focuses on the islanded μG including thermal power plant with 20 MW, wind power with 8 MW, solar power with 4 MW, domestic loads with 15 MW, and ESS with 5 MW. The system base is 20 MW. The simplified model of the islanded

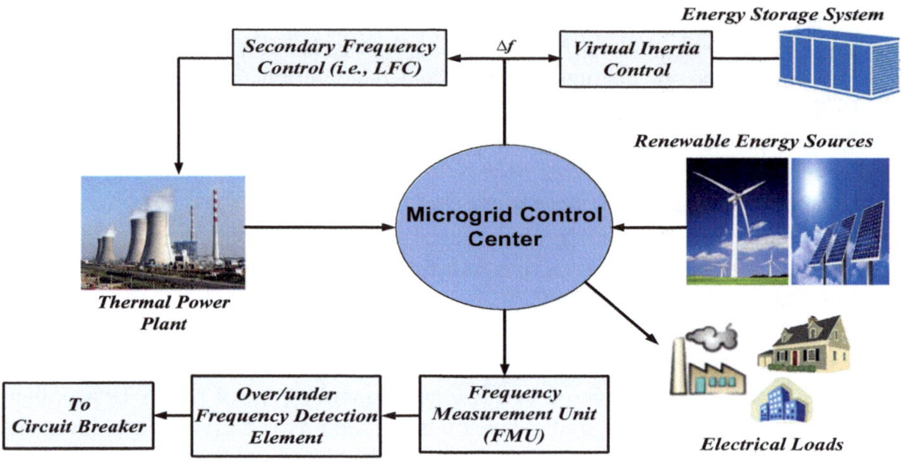

Fig. 4.1 A simplified model of the islanded μG with the proposed coordination scheme [102]

μG is shown in Fig. 4.1. To obtain an accurate perception of the studied μG, this study takes into consideration the important inherent nonlinearity requirements and the physical constraints enjoined by the system dynamics of the generation units. One of the most important constraints of power plants is the rate of generation power change because of the limitation of mechanical movements. Dynamics of the physical system of power plants is represented by GRCs, which limit the rate of the generation output power, and the limit of the valve gate (i.e., the maximum/minimum limit). The GRC for the non-reheat thermal power plant is specified as 20% pu MW/min [8]. The maximum and minimum limits (V_U, V_L) restrict the valve opening/closing. Therefore, the dynamic model of the islanded μG with the proposed coordination scheme is shown in Fig. 4.2. The dynamic parameters of the studied μG considering RESs are given in Table 4.1.

4.2.2 *Virtual Inertia Control for μGs*

The RESs can bring significant impacts to the system inertia ($H_{\mu G}$) when increasing their penetration level. The RESs exchange power to the μGs through power electronic devices (i.e., inverters/converters). These power converter interface-based RESs are static devices without any rotating mass so that the associated inertia constant is roughly zero [3]. Hence, the overall system inertia will be reduced along with increasing the penetration level of RESs, which leads to an increase in the system frequency deviation. Moreover, such low inertia constant renders the μGs more vulnerable to the transients than the bulk power systems. To overcome this problem, the virtual inertia control, which is a special case of VSG execution is implemented. As mentioned before, the activity of the prime mover is imitated to

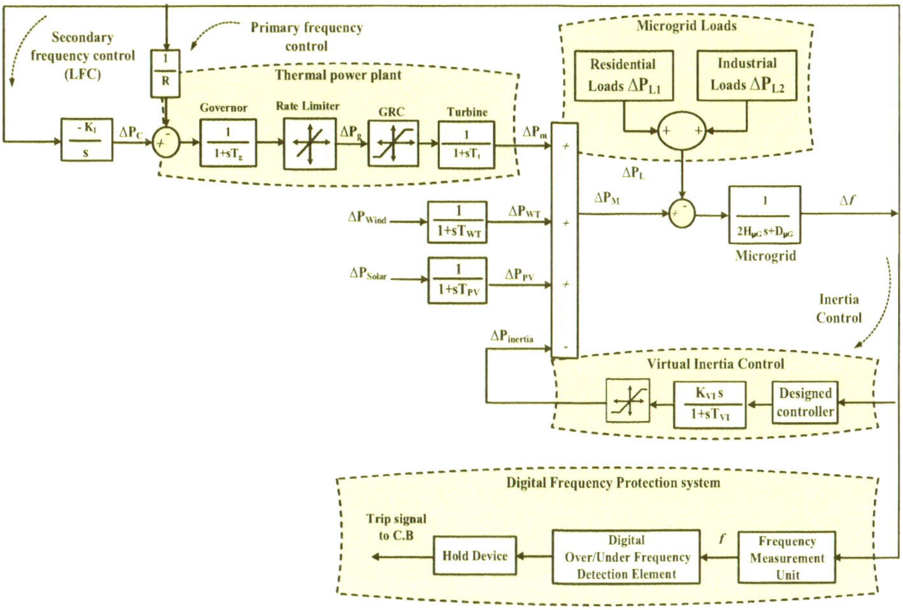

Fig. 4.2 A dynamic model of the studied islanded μG with the proposed coordination scheme

Table 4.1 Dynamic parameters of the islanded μG

Parameter	Value	Parameter	Value
$D_{\mu G}$	0.015	K_I	0.05
$H_{\mu G}$	0.083	K_{VI}	0.5
T_g	0.1	T_{VI}	10
T_t	0.4	V_U	0.3
T_{WT}	1.5	V_L	−0.3
T_{PV}	1.8	GRC	20%
R	2.4	f	50

support frequency stability of the islanded μG [45]. Therefore, lack of inertia response resulting from the RESs in the islanded μG can be compensated by adding active power to the set point, which is simulated by the block of virtual inertia control as shown in Fig. 4.2. The RoCoF function is used for many applications such as fast load shedding, to speed up operation time in over/underfrequency situations and to detect loss of grid. Therefore, the virtual inertia control strategy is based on RoCoF where it calculates the deviation of system frequency to add a compensation active power to the set point of the islanded μG during high-level RES penetration and contingencies [44, 104]. The RoCoF is calculated using the derivative control that is the main concept of the virtual inertia control as follows:

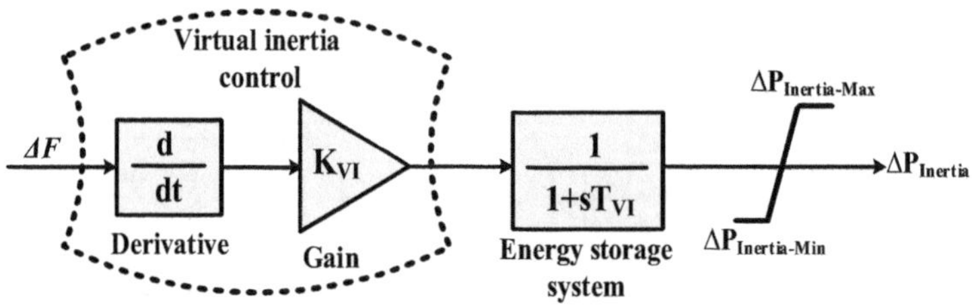

Fig. 4.3 A dynamic structure of the designed virtual inertia control

$$\text{RoCoF} = \left[\frac{\mathrm{d}(\Delta f)}{\mathrm{d}t}\right] \tag{4.1}$$

The control strategy law for imitating inertia power in the form of Laplace based on the per-unit system is shown in Fig. 4.3. The proposed virtual inertia control system gives the desired power to the μG during the deviation of frequency and essentially when the RESs are highly penetrated as follows [43, 44]:

$$\Delta P_{\text{Inertia}} = \frac{K_{\text{VI}}}{1 + sT_{\text{VI}}}\left[\frac{\mathrm{d}(\Delta f)}{\mathrm{d}t}\right] \tag{4.2}$$

where T_{VI} is the time constant-based virtual inertia to emulate the dynamic control of the ESS in the islanded μG, and K_{VI} is the gain of virtual inertia control in the islanded μG. The criteria for selecting these two parameters are related to the stability of the μG and the response required in dynamics. Moreover, it the initial RoCoF as well as the maximum drop in frequency (i.e., nadir frequency) must be taken into consideration.

4.3 State-Space Dynamic Modeling

The state-space model is an effective model for the islanded μG, which is the sixth-order linearized power system considering high-penetration RESs. The deviation of the frequency of the studied islanded μG considering the effect of the primary control loop (i.e., governor action), a secondary control loop (i.e., LFC), and supplementary control loop (i.e., virtual inertia control) can be obtained as

$$\dot{\Delta f} = \frac{1}{2H_{\mu\text{G}}}(\Delta P_{\text{m}} + \Delta P_{\text{WT}} + \Delta P_{\text{PV}} \pm \Delta P_{\text{Inertia}} - \Delta P_{\text{L}}) - \frac{D_{\mu\text{G}}}{2H} \times \Delta f \tag{4.3}$$

where

$$\Delta \dot{P}_{\mathrm{g}} = -\frac{1}{T_{\mathrm{g}}}(\Delta P_{\mathrm{g}}) - \frac{1}{R \cdot T_{\mathrm{g}}} \times \Delta f + \frac{1}{T_{\mathrm{g}}}(\Delta P_{\mathrm{C}}) \tag{4.4}$$

$$\Delta \dot{P}_{\mathrm{m}} = -\frac{1}{T_{\mathrm{t}}}(\Delta P_{\mathrm{m}}) + \frac{1}{T_{\mathrm{t}}}(\Delta P_{\mathrm{g}}) \tag{4.5}$$

$$\Delta \dot{P}_{\mathrm{WT}} = \frac{1}{T_{\mathrm{WT}}}(\Delta P_{\mathrm{Wind}}) - \frac{1}{T_{\mathrm{WT}}}(\Delta P_{\mathrm{WT}}) \tag{4.6}$$

$$\Delta \dot{P}_{\mathrm{PV}} = \frac{1}{T_{\mathrm{PV}}}(\Delta P_{\mathrm{Solar}}) - \frac{1}{T_{\mathrm{PV}}}(\Delta P_{\mathrm{PV}}) \tag{4.7}$$

$$\Delta \dot{P}_{\mathrm{Inertia}} = -\frac{1}{T_{\mathrm{VI}}}(\Delta P_{\mathrm{Inertia}}) + \frac{K_{\mathrm{VI}}}{T_{\mathrm{VI}}}(\Delta \dot{f}) \tag{4.8}$$

In this study, the wind power variation (ΔP_{Wind}), the solar power variation ($\Delta P_{\mathrm{Solar}}$), and the load power variation (ΔP_{L}) are considered as the system disturbance signals. Using suitable definitions and state variables from Eqs. (4.3)–(4.8), the linearized state-space model of the islanded μG can be easily obtained in the form of the following equations:

$$\dot{X} = \mathrm{AX} + \mathrm{BU} + \mathrm{EW} \tag{4.9}$$

$$Y = \mathrm{CX} + \mathrm{DU} + \mathrm{FW} \tag{4.10}$$

where

$$X^T = [\Delta f \;\; \Delta P_{\mathrm{g}} \;\; \Delta P_{\mathrm{m}} \;\; \Delta P_{\mathrm{WT}} \;\; \Delta P_{\mathrm{PV}} \;\; \Delta P_{\mathrm{Inertia}}]$$

$$W^T = [\Delta P_{\mathrm{Wind}} \Delta P_{\mathrm{Solar}} \Delta P_{\mathrm{L}}]$$

$$Y = [\Delta f]$$

where U is the control input signal, W is the input disturbance vector, X is the state vector, and Y is the control output signal, which is the system frequency deviation (Δf). While B corresponds to the control input signal, E corresponds to the disturbance inputs, D is the zero vector with the same size of the input control signal, F is the zero vector with the same size of the input disturbance vector, and C corresponds to the output measurement, which is the input to load frequency controller. Thus, the complete equations of the state-space representation of the islanded μG considering RESs can be obtained as in Eq. (4.11):

$$\dot{X}=\begin{bmatrix} -\frac{D_{\mu G}}{2H_{\mu G}} & 0 & \frac{1}{2H_{\mu G}} & \frac{1}{2H_{\mu G}} & \frac{1}{2H_{\mu G}} & \frac{1}{2H_{\mu G}} \\ -\frac{1}{RT_g} & -\frac{1}{T_g} & 0 & 0 & 0 & 0 \\ 0 & \frac{1}{T_t} & -\frac{1}{T_t} & 0 & 0 & 0 \\ 0 & 0 & 0 & -\frac{1}{T_{WT}} & 0 & 0 \\ 0 & 0 & 0 & 0 & -\frac{1}{T_{PV}} & 0 \\ \frac{-K_{VI}*D_{\mu G}}{T_{VI}*2H_{\mu G}} & 0 & \frac{K_{VI}}{T_{VI}*2H_{\mu G}} & \frac{K_{VI}}{T_{VI}*2H_{\mu G}} & \frac{K_{VI}}{T_{VI}*2H_{\mu G}} & \left(\frac{K_{VI}}{T_{VI}*2H_{\mu G}}-\frac{1}{T_{VI}}\right) \end{bmatrix} * \begin{bmatrix} \Delta f \\ \Delta P_g \\ \Delta p_m \\ \Delta p_{WT} \\ \Delta P_{PV} \\ \Delta P_{Inertia} \end{bmatrix}$$

$$+\begin{bmatrix} 0 \\ \frac{1}{T_g} \\ 0 \\ 0 \\ 0 \\ 0 \\ 0 \end{bmatrix} *[\Delta P_C]+\begin{bmatrix} 0 & 0 & -\frac{1}{2H_{\mu G}} \\ 0 & 0 & 0 \\ 0 & 0 & 0 \\ \frac{1}{T_{WT}} & 0 & 0 \\ 0 & \frac{1}{T_{PV}} & 0 \\ 0 & 0 & -\frac{K_{VI}}{T_{VI}*2H_{\mu G}} \end{bmatrix} * \begin{bmatrix} \Delta P_{Wind} \\ \Delta P_{Solar} \\ \Delta P_L \end{bmatrix}$$

$$Y=[1\ 0\ 0\ 0\ 0\ 0]*X \tag{4.11}$$

4.4 Virtual Inertia Control Based on the Optimal PI Controller

The system inertia of the μG has a key role in system robustness concerning frequency stability; this arises due to an imbalance between the generation and load demand. In conventional power systems, conventional SGs are considered as the main source of the dynamics. Here the SGs play a key role in the limitation of the RoCoF as well as give a natural response to system frequency deviations in the wake of contingencies such as an unplanned generation loss or load in the power system [44]. On the other hand, in μGs, the RESs are considered as the host of dynamics. Therefore, utilizing the power electronic inverter-based RESs reduces overall system inertia and affects system stability and reliability. Hence, to create sufficient virtual inertia-based inverters, this study proposes the virtual inertia controller based on the optimal PI controller to emulate inertia characteristic, which contributes to the total inertia of the islanded μG, thus enhancing the frequency stability and resilience.

Moreover, the PSO algorithm is applied to minimize the system frequency deviations and yields the optimal design parameters of the PI controller under the operating conditions of the islanded μG.

4.4.1 Optimal PI Controller Design for Virtual Inertia Control

The main target of this section is to design the optimal PI controller that produces the optimal control signal to the virtual inertia control loop for emulating the inertia response into the μG, thus improving frequency performance and stability, and avoiding system collapse during contingencies. The PI controller is composed of proportional gain K_p and integral gain K_i. Its transfer function is expressed as follows:

$$G(s) = K_\mathrm{p} + \frac{K_\mathrm{i}}{s} \tag{4.12}$$

An intelligent searching method (i.e., the PSO) is used to find the optimal design parameters of the PI controller for minimizing the system frequency deviation (i.e., objective function). In the literature, the ISE criterion has been used as the objective function for optimum tuning of PI/PID controller's parameters in several optimization techniques [83, 109]. Therefore, in this study, the objective function of the optimization problem depends on the ISE criterion, which can be formulated as follows:

$$\mathrm{ISE} = \int_0^{t_\mathrm{sim}} (\Delta f)^2 \mathrm{d}t \tag{4.13}$$

This is subject to bounds of the PI controller parameters as [83]

$$\left[K_{p,i}^{\mathrm{Min}} \le K_{p,i} \le K_{p,i}^{\mathrm{Max}}\right]$$

where (Δf) is the frequency deviation of the islanded μG and t_sim is the simulation time to execute one run. The proposed PSO technique is applied in the μG to obtain the minimum value of the objective function (i.e., system frequency deviation) through getting the optimal PI controller parameters.

4.4.2 Particle Swarm Optimization

The PSO is a global optimization algorithm based on evolutionary computation technique. It was presented by Kennedy and Eberhart in 1995 [110]. The PSO possesses some merits over other optimization algorithms (i.e., genetic algorithm, artificial neural networks, fuzzy logic, and ant colony) for tuning PI/PID controller parameters such as the ease of use, high convergence rates, minimum storage requirements, and less depending on the set of initial values, implying the robustness [111]. Considering these advantages, this study proposes the PSO algorithm to tune the PI controller parameters, obtaining the optimum PI controller parameters with the robustness of operations. The basic process principle of this optimization method is developed on a flock of flocking birds. Birds are either scattered or go from place to place to look for food. Furthermore, one can discover where food can be found because it is transported by other birds at any time during the search for food [112]. In this optimization technique, rather than utilizing evolutionary operators, individuals called particle are used. In this way, a swarm comprises a number of particles, every particle representing the possibility of the problem. Every particle in the PSO algorithm flies through the search space according to its own particular flying knowledge and its friend flying background. Each particle is dealt with as a particle in the D-dimension search space. The location of the particle is represented as X_i; the best previous mode of any particle is recorded and called the P_{best}. Another best value (i.e., the overall best value g_{best}) is followed by a global version of the PSO algorithm [113]. The particle velocity i is represented by V_i and all particles are updated as per the following equations:

$$v_{\text{id}}^{n+1} = w \cdot V_{\text{id}}^{n} + c_1 \cdot \text{rand}\ () \cdot \left(P_{\text{id}}^{n} - X_{\text{id}}^{n}\right) + c_2 \cdot \text{rand}\ () \cdot \left(P_{\text{gd}}^{n} - X_{\text{id}}^{n}\right) \tag{4.14}$$

$$x_{\text{id}}^{n+1} = x_{\text{id}}^{n} + v_{\text{id}}^{n+1} \tag{4.15}$$

These equations are utilized to estimate the new values of velocity and position of every particle as per its previous values. Learning factors of the optimization technique have significant implications on the algorithm convergence rate. Further information for the PSO can be found in [110–113]. In this study, the main objective of the PSO algorithm is to minimize the objective function as given in Eq. (4.13) through finding the optimal parameters of the PI controller. Therefore, the diagram of the proposed PSO algorithm for optimum parameters (i.e., K_{p} and K_{i}) of the PI controller is illustrated in Fig. 4.4. The performance of the PSO in searching the PI controller parameters in the studied μG has been validated by using the characteristics of the PSO; the number of iteration (n) is 50, number of particles in a group is 50, inertia weight factor (w) is 0.8, and acceleration constants (c_1 and c_2) are 0.12 and 2, respectively. Afterward, the designed optimal PI controller for emulating the inertia response has been achieved by using the PSO. Hence, the optimal PI controller parameters are $K_{\text{p}} = 3.4957$ and $K_{\text{i}} = 22.0991$, which produce the optimal control signal to the virtual inertia control loop for emulating the inertia response into the μG during high RES penetration.

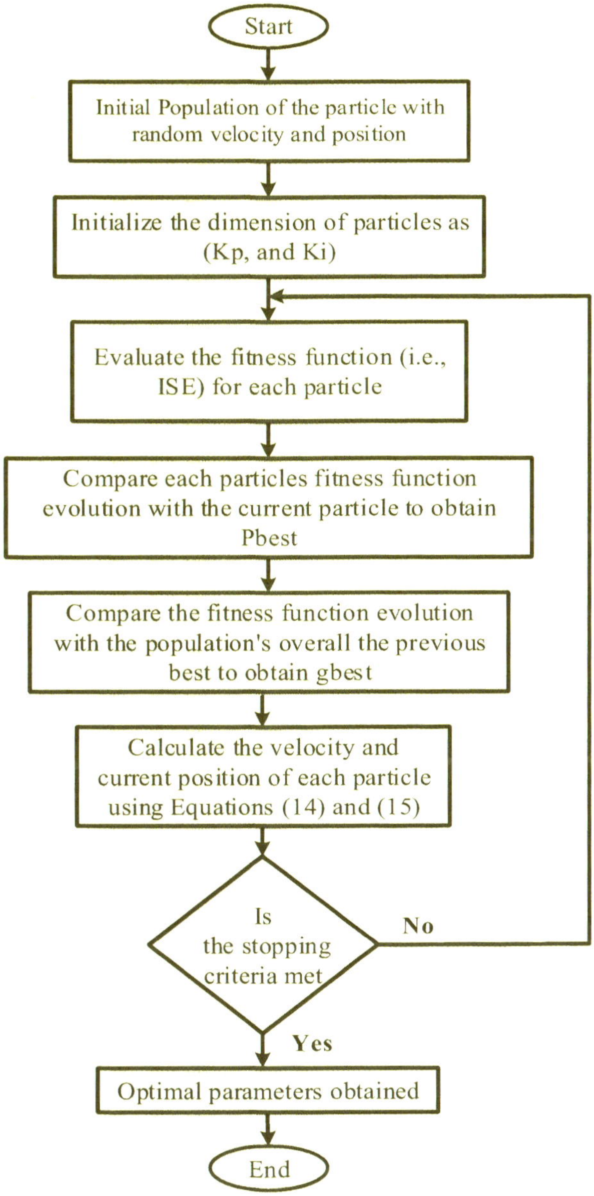

Fig. 4.4 Flowchart of the PSO algorithm [102]

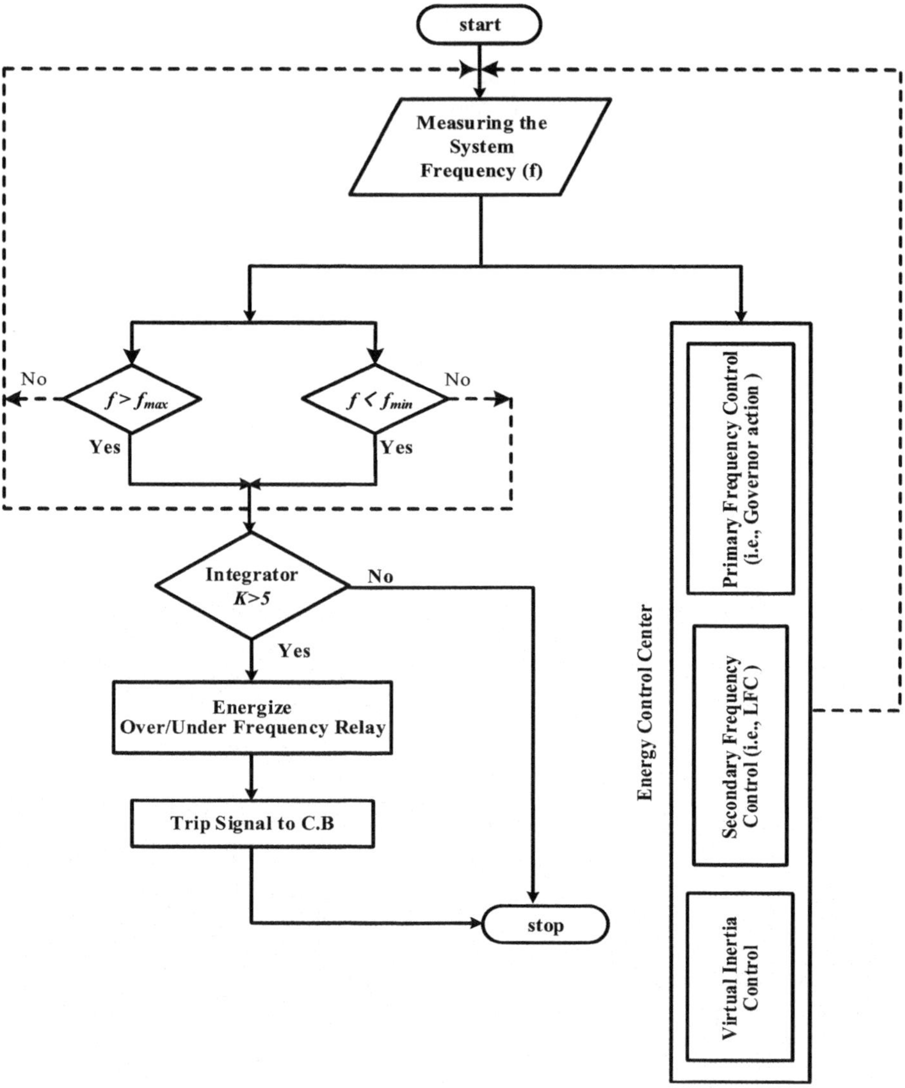

Fig. 4.5 The diagram of the proposed coordination strategy [102]

4.5 Modeling of Digital Frequency Relay

The used model of the digital OUFR has already been described in Chap. 3. Moreover, the principal operation of the OUFR has also been explained in Chap. 3. Hence, the operating sequence of the proposed coordinated scheme of frequency control including virtual inertia control and digital OUFR for preservation of the μG dynamic security is illustrated in Fig. 4.5.

4.6 Simulation Results and Discussions

The simulation results of the studied islanded μG are carried out using MATLAB/ Simulink® software to validate the effectiveness and robustness of the proposed coordination scheme under the nature variety of the RESs, random load variation, and system inertia variations (i.e., system uncertainties). The code of the PSO as an m-file is interfaced with the model of the islanded μG to execute the optimization process. Preservation of the μG dynamic security with the proposed coordinated scheme is investigated under different operating conditions through the following scenarios:

4.6.1 Performance Evaluation of the μG Without the RESs

4.6.1.1 Scenario A: Multiple Disturbances in Load Demand

The main target of this scenario is to investigate the performance of the islanded μG with the proposed coordinated scheme by implementing high-fluctuated random load variation (i.e., industrial loads) and low-fluctuated random load variation (i.e., residential load) as shown in Fig. 4.6. This scenario is divided into three sub-scenarios, which are implemented under the assumed various operating states of random load variants in Table 4.2.

Scenario 1A: In this case, it is assumed that the default parameters of the μG system studied are 100% of the system's default inertia as well as the multiple operating conditions of random load variation are implemented as shown in Table 4.2. The frequency response of the studied μG is shown in Fig. 4.7, where the μG frequency fluctuates within the limits of the digital relay. Therefore, in this case, the role of the frequency control loops emerges without any action from the digital OUFR. From Fig. 4.7, it is clear that the frequency change of the μG without virtual inertia controller is ±0.75 Hz, while the studied μG with the dynamic contribution of virtual inertia controller gives a frequency change of ±0.63 Hz during the connection of the industrial load (i.e., high-fluctuated random load variation). Compared to the μG with/without the virtual inertia controller, the proposed coordination scheme of frequency control including virtual inertia controller based on the optimal PI controller and the digital OUFR can give smooth and secure frequency performance, about ±0.3 Hz, when the high-fluctuated load is connected at $t = 300$ s. Hence, the preservation of the studied μG dynamic security has been achieved by using the proposed virtual inertia controller based on the optimal PI controller.

Scenario 2A: In this case, the μG system studied is assumed to have the default system parameters and the μG is examined under the situation of 50% of system's default inertia with the multiple operating conditions of random load variation as shown in Table 4.2. The frequency response of the studied islanded μG is shown in

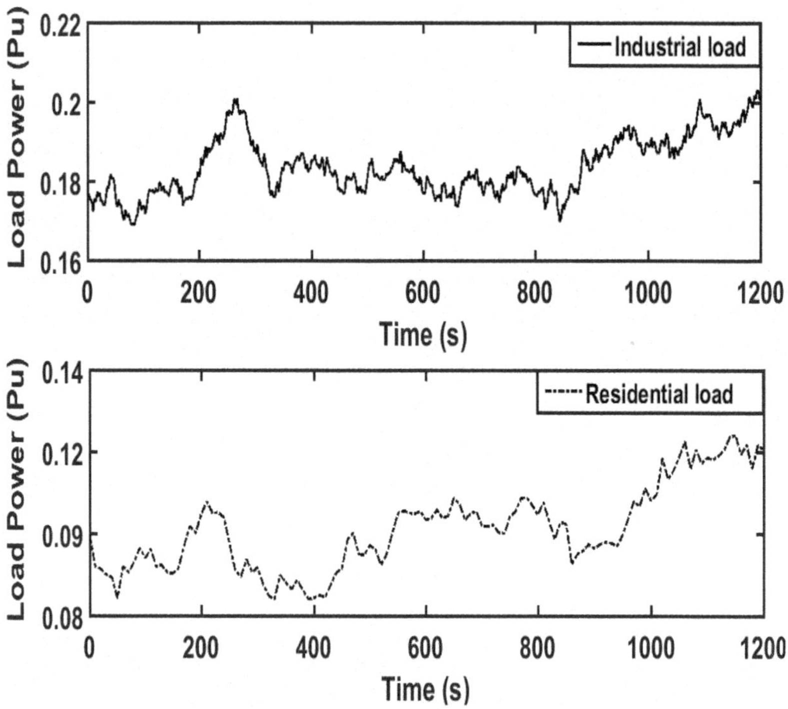

Fig. 4.6 The random load deviation of the case studied

Table 4.2 Multiple operating conditions of the islanded μG for Scenario 1

Domestic loads	Starting time (s)	Stopping time (s)	Size (pu)
Industrial load	300 s	–	0.21
Residential load	Initial	800 s	0.12

Fig. 4.8. The system response of the μG without virtual inertia controller fluctuates beyond the permissible frequency limits when the industrial load is connected at $t = 300$ s as well as the integrator output magnitude is larger than the integrator set time ($K = 5$ s). Thus, the digital OUFR energizes and sends a trip signal to the generator circuit breaker. Hence, the action of the digital OUFR emerges at this stage. On the other hand, although the system response of the μG with virtual inertia controller fluctuates beyond the permissible frequency ranges at $t = 300$ s, the digital OUFR does not energize as the integrator output magnitude does not exceed the specified value. And then, the frequency control loops can readjust the frequency to its normal value and maintain the μG dynamic security. The best frequency performance is obtained by utilizing the proposed virtual inertia controller based on the optimal PI controller, which can maintain the system frequency change within ± 0.38 Hz during the connection of the industrial load.

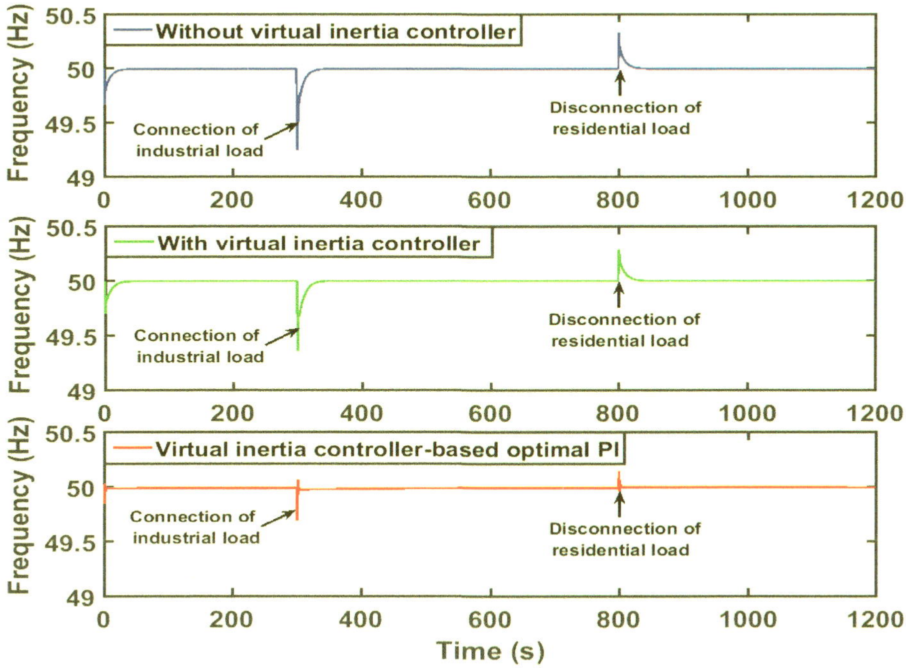

Fig. 4.7 The frequency response of the studied μG for Scenario 1A (i.e., 100% of default system inertia)

Scenario 3A: The effectiveness and robustness of the proposed coordination scheme of protection and frequency stability are tested in the extreme case where the studied μG system is examined under the situation of 30% of default system inertia with the various operating conditions of random load variation as shown in Table 4.2. The frequency response of the studied μG with different control strategies' incorporation with the digital OUFR is shown in Fig. 4.9. In the case of the studied μG without virtual inertia controller, the frequency control loops cannot endure the variation of μG frequency due to connecting of the residential load at $t = 0$ s, so the digital OUFR energizes and sends a trip signal to the generator circuit breaker at that time. The system response of the μG with virtual inertia controller fluctuates beyond the allowable frequency limits when the high-fluctuated load is connected at $t = 300$ s. However, the integrator output does not overtake the integrator set time K. Thus, the virtual inertia controller besides the frequency control loops can restore the frequency to the nominal value at 50 Hz. On the other hand, the studied μG with the proposed virtual inertia control does not need action from the digital OUFR; the proposed virtual inertia control succeeded to maintain the μG frequency to its nominal value. Hence, the proposed coordination effectiveness was approved to maintain the μG dynamic security.

The performance specifications, maximum overshoot (MOS), maximum undershoot (MUS), and maximum settling time (T_S) of the studied μG, with different

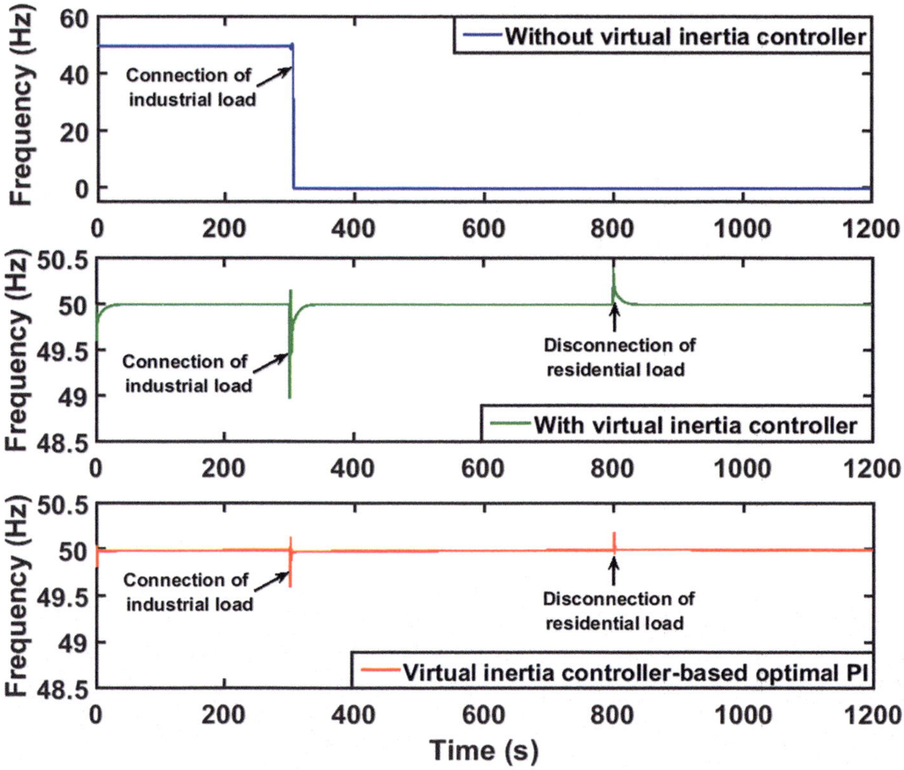

Fig. 4.8 The frequency response of the studied μG for Scenario 2A (i.e., 50% of default system inertia)

control strategies (i.e., virtual inertia control based on the optimal PI controller, virtual inertia controller, and without virtual inertia controller) for Scenario A during the whole period of simulation (20 min) have been compared in Table 4.3.

4.6.1.2 Scenario B: Abrupt Load Change

In this scenario, the performance of the islanded μG with the proposed coordination scheme is tested and evaluated by using 2.0 MW step load change ($\Delta P_L = 0.1$ pu) at time $t = 200$ s. Moreover, this scenario is divided into three sub-scenarios, which are implemented under the variations of system inertia (i.e., system uncertainties).

Scenario 1B: In this case, it is assumed that the default parameters of the studied system are 100% of the system's default inertia as well as a 10% pu step change at time $t = 200$ s is implemented. Figure 4.10 displays the frequency response of the studied μG with the different control strategies. The μG frequency fluctuates within the allowable limits of the system frequency. From Fig. 4.10, it is obvious that the studied μG with the proposed coordination scheme of frequency control including

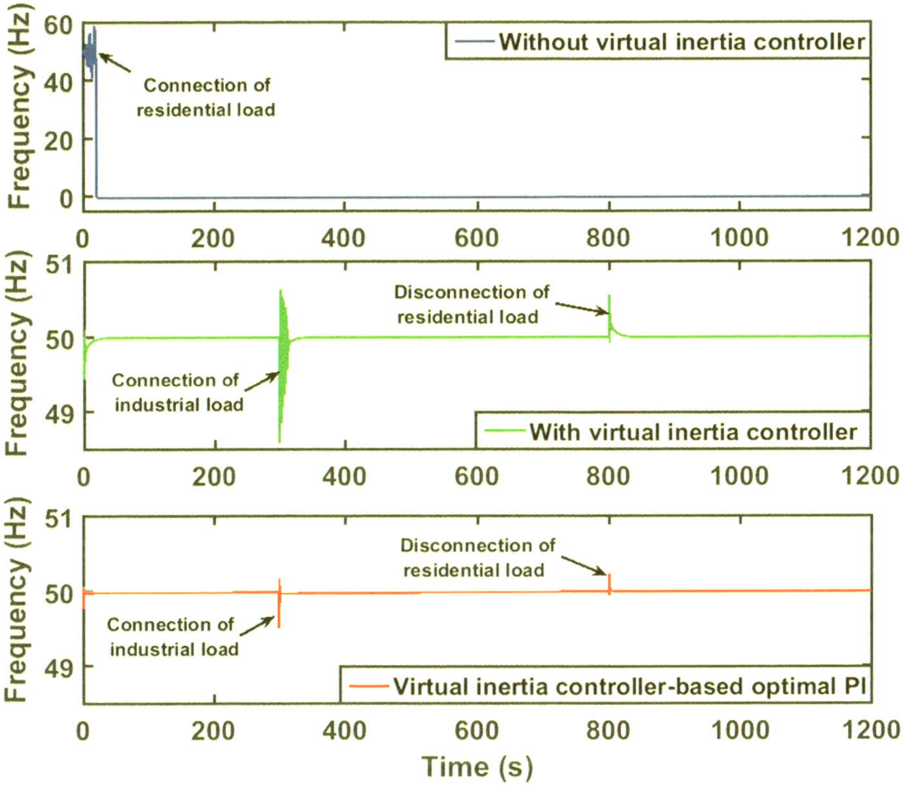

Fig. 4.9 The frequency response of the studied μG for Scenario 3A (i.e., 30% of default system inertia)

Table 4.3 The performance specifications of the studied μG for Scenario A

Scenario A	Virtual inertia controller-based optimal PI controller			With virtual inertia controller			Without virtual inertia controller		
	MUS (pu)	MOS (pu)	T_S (s)	MUS (pu)	MOS (pu)	T_S (s)	MUS (pu)	MOS (pu)	T_S (s)
High system inertia (100%)	0.300	0.146	5.0	0.630	0.292	23.0	0.757	0.330	30.0
Medium system inertia (50%)	0.380	0.190	6.0	1.000	0.405	24.0	–	–	–
Low system inertia (30%)	0.457	0.220	6.0	1.105	0.550	26.0	–	–	–

virtual inertia controller-based optimal PI controller and digital OUFR is more stable and faster, compared to that with/without virtual inertia controller.

Scenario 2B: In this case, the μG frequency response under the situation of medium system inertia (i.e., 50% of its value) is examined. Moreover, a 10% pu of

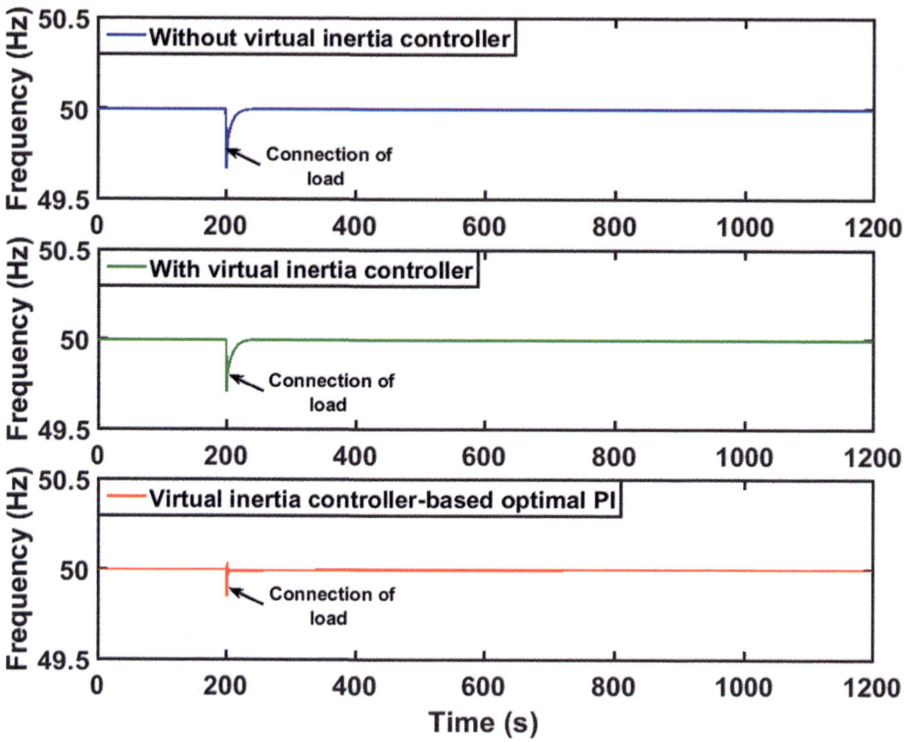

Fig. 4.10 The frequency response of the studied μG for Scenario 1B (i.e., 100% of default system inertia)

step load change is implemented to the studied μG at time $t = 200$ s. The frequency response of the studied μG with the proposed control strategies' incorporation with the digital OUFR is shown in Fig. 4.11. From Fig. 4.11, it is clear that the virtual inertia control-based optimal PI controller, which is optimally designed using PSO algorithm, significantly enhances frequency performance and decreases the magnitude of system transients, compared to the studied μG with/without the virtual inertia controller.

Scenario 3B: To perform a more drastic simulation, the studied μG is investigated under low system inertia (i.e., 30% of its nominal values) as well as a 10% pu step change at time $t = 200$ s is implemented. Figure 4.12 shows the frequency response of the studied μG with the different control strategies (i.e., virtual inertia control-based optimal PI controller, virtual inertia controller, and without virtual inertia controller) incorporating with the digital OUFR. The frequency response of the μG without virtual inertia controller fluctuates beyond the allowable frequency limits when a step load change is connected at $t = 200$ s as well as the integrator output magnitude is larger than the integrator set time ($K = 5$ s). Thus, the digital OUFR energizes and sends a trip signal to the generator circuit breaker. On the other hand, the frequency control loops can readjust the frequency to its normal value and

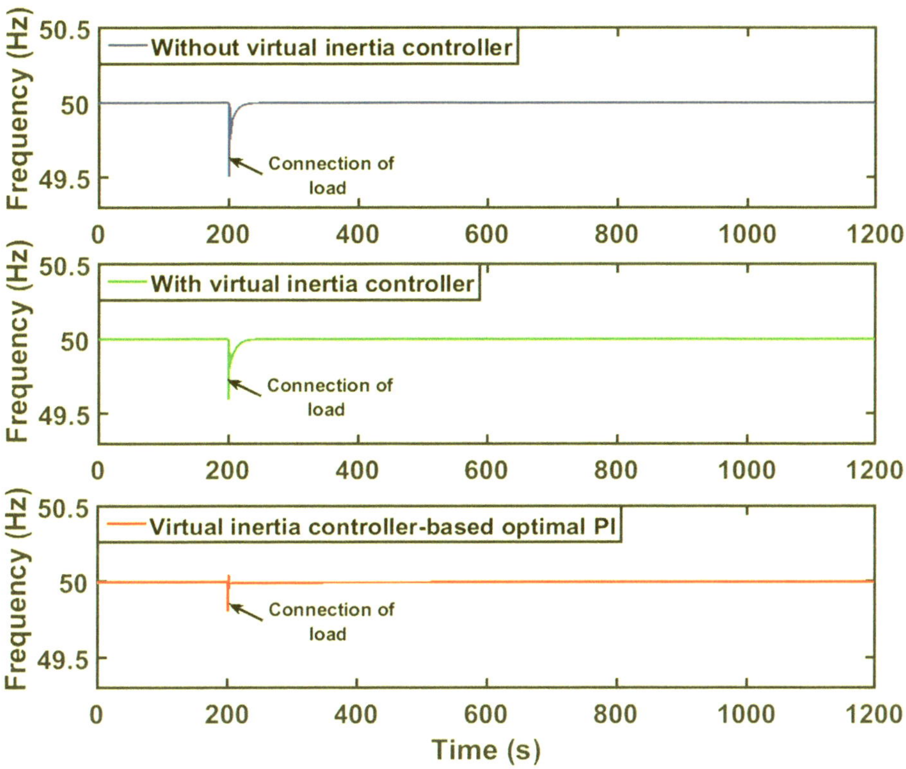

Fig. 4.11 The frequency response of the studied μG for Scenario 2B (i.e., 50% of default system inertia)

maintain the μG dynamic security in case of the studied μG with/without virtual inertia controller. However, the studied μG with the proposed virtual inertia controller-based optimal PI controller is more stable and faster, compared to that with the virtual inertia controller.

The performance specifications, (MOS), (MUS), and (T_S) of the studied μG, with different control strategies (i.e., virtual inertia control-based the optimal PI controller, virtual inertia controller, and without virtual inertia controller) for Scenario B during the whole period of simulation (20 min) have been compared in Table 4.4.

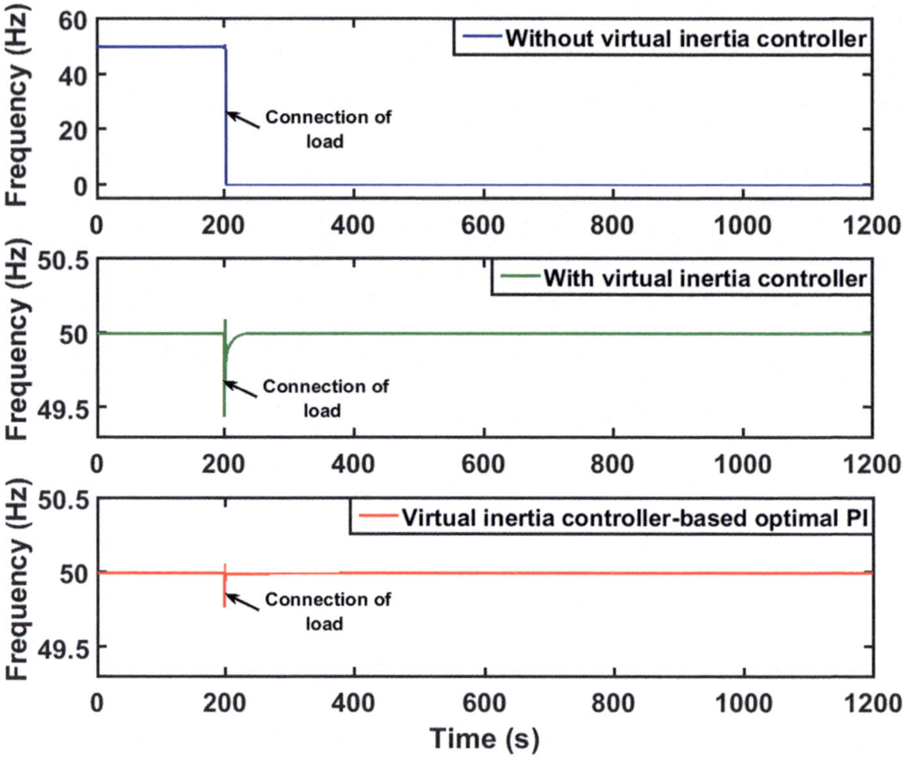

Fig. 4.12 The frequency response of the studied μG for Scenario 3B (i.e., 30% of default system inertia)

Table 4.4 The performance specifications of the studied μG for Scenario B

	Virtual inertia controller-based optimal PI controller			With virtual inertia controller			Without virtual inertia controller		
Scenario B	MUS (pu)	MOS (pu)	T_S (s)	MUS (pu)	MOS (pu)	T_S (s)	MUS (pu)	MOS (pu)	T_S (s)
High system inertia (100%)	0.145	0.003	4.0	0.2928	0.0002	21.0	0.3312	0.00005	28.0
Medium system inertia (50%)	0.184	0.004	6.0	0.3991	0.0001	22.0	0.5001	0.00005	30.0
Low system inertia (30%)	0.231	0.060	7.0	0.5610	0.0952	25.0	–	–	–

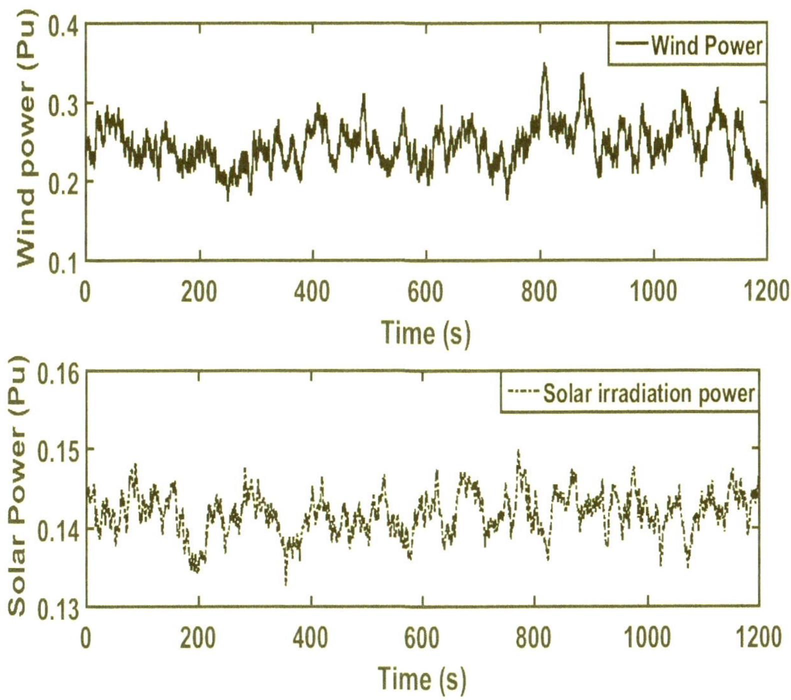

Fig. 4.13 Power variation patterns of wind and solar irradiation power

4.6.2 Performance Evaluation of the μG Including the RES Uncertainty

4.6.2.1 Scenario C: Multiple Disturbances in Load Demand

The main target of this scenario is to evaluate the performance of the islanded μG with the proposed coordinated scheme under variation in loading conditions, system inertia (i.e., system uncertainties), and nature of the RESs, which are defined as the important properties of the actual μG. Therefore, the islanded μG studied is tested by implementing high-fluctuated wind power and low-fluctuated solar power as shown in Fig. 4.13 as well as implementating industrial and residential loads as shown in Fig. 4.6. This scenario is divided into three sub-scenarios, which are implemented under the assumed various operating states as shown in Table 4.5.

Scenario 1C: In this case, the islanded μG studied is examined under the situation of 100% of default system inertia as well as by implementing the assumed various operating conditions as shown in Table 4.5. Figure 4.14 shows the response of the μG frequency, where the μG frequency fluctuates within the allowable frequency limits. Therefore, in this case, the role of the frequency control loops emerges without any action from the digital OUFR. From Fig. 4.14, it can be seen that the frequency change of the μG without virtual inertia controller is ±0.95 Hz while the

Table 4.5 Multiple operating conditions of the islanded μG for Scenario 2

Disturbance source	Starting time (s)	Stopping time (s)	Size (pu)
Industrial load	300 s	–	0.21
Residential load	Initial	800 s	0.12
Wind farm	600 s	–	0.35
Solar power plant	Initial	–	0.14

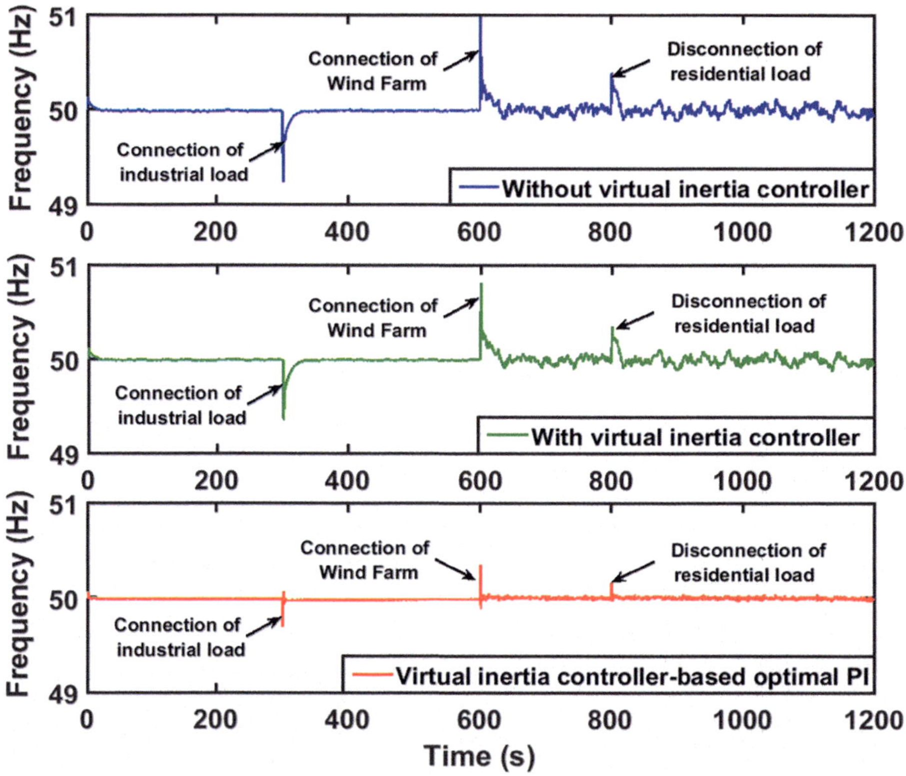

Fig. 4.14 The frequency response of the studied μG for Scenario 1C (i.e., 100% of default system inertia)

studied μG with the dynamic contribution of virtual inertia controller gives a frequency change of ±0.82 Hz when the wind farm (i.e., high-fluctuated wind power) is connected at $t = 600$ s. On the other hand, the proposed virtual inertia controller-based optimal PI controller enhances the frequency performance (i.e., the frequency change of about ±0.32 Hz) and reduces the magnitude of system transients compared to the studied μG with/without the conventional virtual inertia controller. Hence, the proposed coordination effectiveness has been validated for maintaining μG dynamic security.

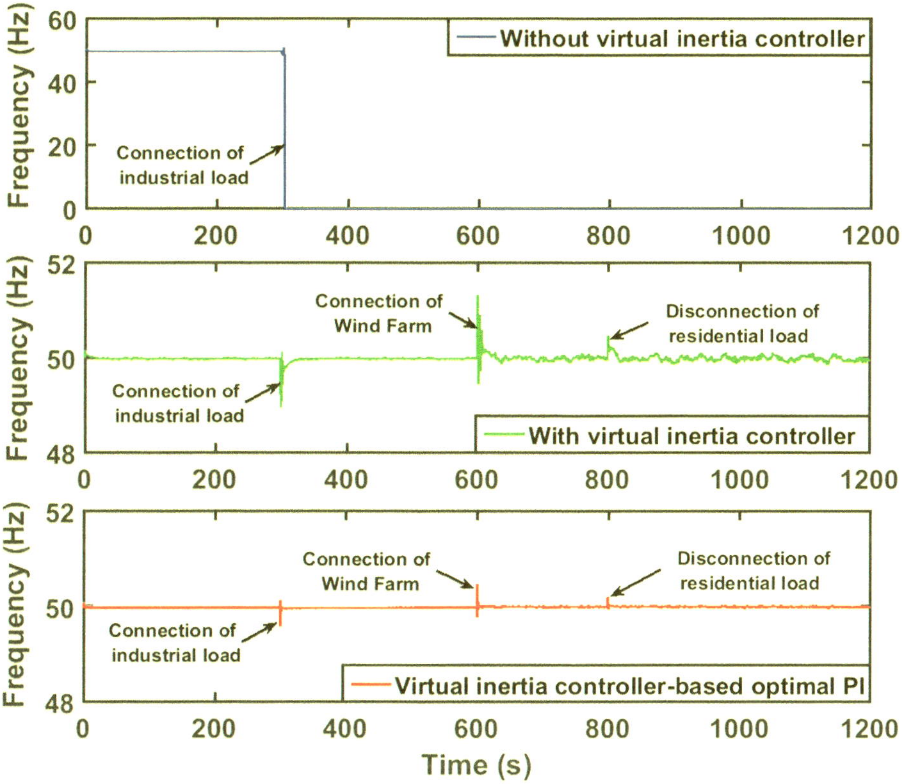

Fig. 4.15 The frequency response of the studied μG for Scenario 2C (i.e., 50% of default system inertia)

Scenario 2C: To perform an extreme test scenario, the studied μG is tested under the situation of reduction of the system inertia to 50% from its initial value with the various operating states as shown in Table 4.5. The response of the studied islanded μG frequency is shown in Fig. 4.15. From Fig. 4.15, it is found that the system response of the studied μG without virtual inertia controller fluctuates severely due to critical inertia reduction, as well as the integrator output of the proposed protection system overtakes the integrator set time K. Thus, the digital OUFR energizes and sends a trip signal to the generator circuit breaker when connecting the industrial load at 300 s. The frequency response of the studied μG with virtual inertia controller fluctuates beyond the allowable frequency limits at critical inertial reduction caused by connecting of wind power at $t = 600$ s. However, the digital OUFR does not energize as the integrator output magnitude does not reach the set value. Hence, the virtual inertia control incorporation with the primary and secondary frequency control can maintain the frequency to its nominal value and maintain the μG dynamic security. On the other hand, the proposed virtual inertia controller-based optimal PI controller can effectively preserve the frequency deviations within the

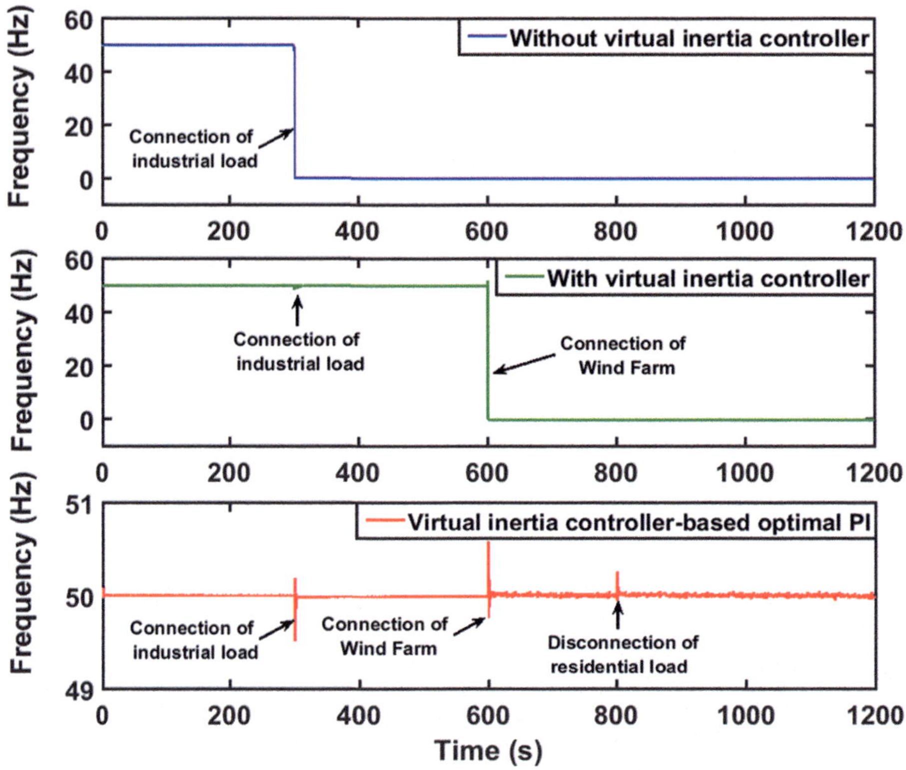

Fig. 4.16 The frequency response of the studied μG for Scenario 3C (i.e., 30% of default system inertia)

permissible frequency operating standard of ±0.43 Hz during connection of wind farm at $t = 600$ s (provided by AEMC [114]).

Scenario 3C: In this case, the studied μG system is examined under a radical change of the operating state; the studied μG is tested under the situation of 30% of default system inertia as well as by implementing the multiple operating conditions as shown in Table 4.5. The μG frequency responses are more volatile with transient increase and deviation following the reduction of system inertia as displayed in Fig. 4.16. From Fig. 4.16, it is found that the digital OUFR energizes and sends a trip signal to the generator circuit breaker in case of the studied μG without virtual inertia controller during the connection of the industrial load at $t = 300$ s. The studied μG with virtual inertia controller cannot withstand the frequency fluctuations during the connection of wind farm and hence the role of the digital OUFR emerges and sends a trip signal to the generator circuit breaker at $t = 600$ s. Compared to the μG with/without the virtual inertia controller, the proposed virtual inertia controller-based optimal PI controller can address the applicable uncertainties and change the frequency rapidly to its nominal value (50 Hz). Furthermore, in this extreme case of

Table 4.6 The performance specifications of the studied μG for Scenario C

Scenario C	Virtual inertia controller-based optimal PI controller			With virtual inertia controller			Without virtual inertia controller		
	MUS (pu)	MOS (pu)	T_S (s)	MUS (pu)	MOS (pu)	T_S (s)	MUS (pu)	MOS (pu)	T_S (s)
High system inertia (100%)	0.2901	0.327	6.0	0.6441	0.8230	19.0	0.7627	0.951	22.0
Medium system inertia (50%)	0.4002	0.431	6.0	1.023	1.198	23.0	–	–	–
Low system inertia (30%)	0.4320	0.487	7.0	–	–	–	–	–	–

uncertainty, the proposed coordination strategy can maintain the dynamic security of the μG and the stable frequency response compared to other comparable methods.

The performance specifications, (MOS), (MUS), and (T_S) of the studied μG, with different control strategies (i.e., virtual inertia control-based optimal PI controller, virtual inertia controller, and without virtual inertia controller) for Scenario C during the whole period of simulation (20 min) have been compared in Table 4.6.

4.6.2.2 Scenario D: Abrupt Load Change

In this scenario, the islanded μG has been considered as a test system to evaluate the proposed coordination scheme of frequency control including a virtual inertia control-based optimal PI controller and the digital frequency relay. The performance of the islanded μG with the proposed coordinated scheme is tested and evaluated by implementing a 10% pu step load change at $t = 200$ s, high-fluctuated wind power at time $t = 600$ s, and low-fluctuated solar power at $t = 0$ s. The power variation patterns of wind and solar power are shown in Fig. 4.13. Moreover, this scenario is divided into three sub-scenarios, which are implemented under the variations of system inertia.

Scenario 1D: In this case, the μG frequency response under the situation of the system's default inertia (i.e., 100% of system inertia) is examined. Although the frequency response is affected by the sudden load change and RES power fluctuations, the μG frequency fluctuates within the allowable frequency limits as shown in Fig. 4.17. Hence, there is no need for the action of the digital frequency relay. Without the virtual inertia controller, the μG frequency has driven to a large frequency fluctuation of about ±0.98 Hz. The virtual inertia controller could maintain the frequency deviation within ±0.761 Hz. The virtual inertia controller-based optimal PI controller could properly maintain the frequency deviation within ±0.33 Hz. It is obvious that the best result is obtained from the optimal PI controller-based virtual inertia controller. Hence, the proposed coordination effectiveness was approved to maintain the μG dynamic security.

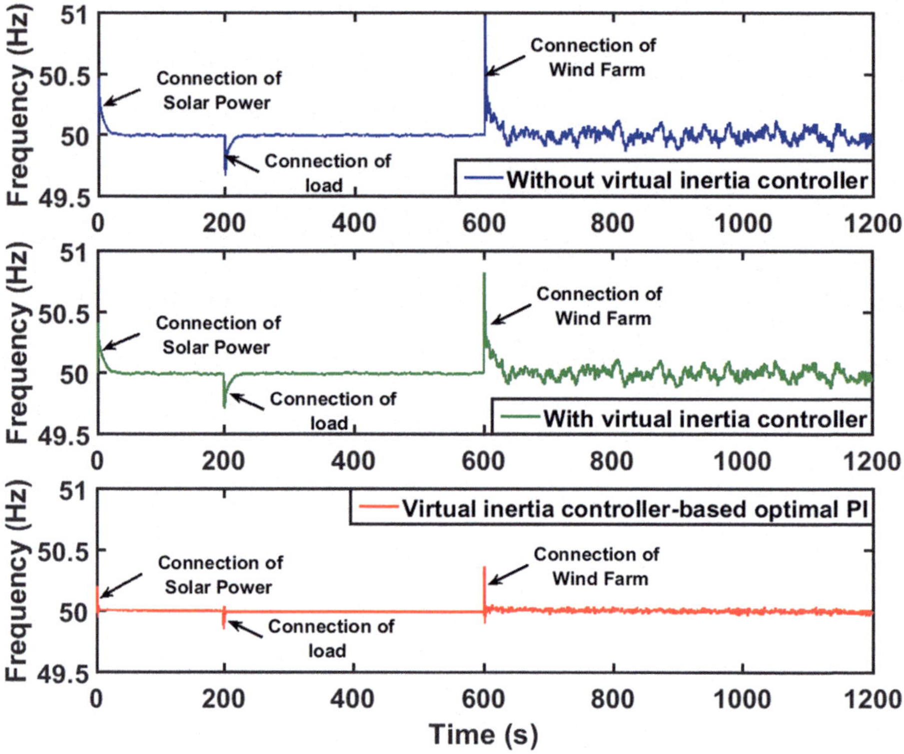

Fig. 4.17 The frequency response of the studied μG for Scenario 1D (i.e., 100% of default system inertia)

Scenario 2D: To make a drastic change in the operation condition, the studied μG is operated under the situation of 50% of system inertia. The μG frequency response is more fluctuating with higher transient and deviation following the reduction in system inertia as shown in Fig. 4.18. From Fig. 4.18, it is found that the frequency response of the studied μG without virtual inertia controller fluctuates severely due to critical inertia reduction, as well as the integrator output of the proposed protection system overtakes the integrator set time *K*. Thus, the digital OUFR energizes and sends a trip signal to the generator circuit breaker when connecting a wind farm at $t = 600$ s. The frequency response of the μG with virtual inertia controller fluctuates beyond the allowable frequency limits when the high-fluctuated wind power is connected at $t = 600$ s. However, the integrator output does not overtake the integrator set time *K*. Thus, the virtual inertia controller can restore the frequency to the nominal value at 50 Hz. On the other hand, the studied μG with the proposed coordination scheme of frequency control including virtual inertia controller-based optimal PI controller and the digital OUFR is more stable and faster, compared to that with virtual inertia controller.

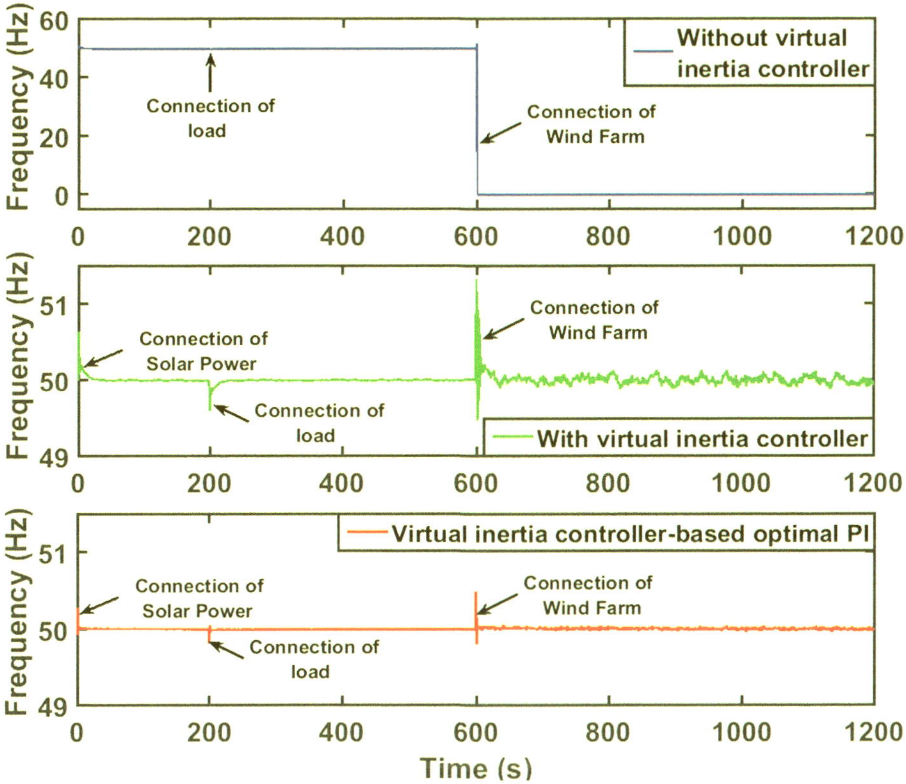

Fig. 4.18 The frequency response of the studied μG for Scenario 2D (i.e., 50% of default system inertia)

Scenario 3D: To carry out an extreme test scenario, the studied μG has been examined under the situation of extremely low system inertia (30% of system inertia). Figure 4.19 displays the frequency response of the studied μG with the different control strategies. In the case of the studied μG without virtual inertia controller, the frequency control loops cannot endure the variation of μG frequency due to connecting of the low-fluctuated solar power at $t = 0$ s, so the digital OUFR energizes and sends a trip signal to the generator circuit breaker at that time. The frequency response of the studied μG with virtual inertia controller fluctuates severely with large deviation during connecting of the high penetration power of wind farm; thus the role of the digital OUFR emerges and sends a trip signal to the generator circuit breaker at $t = 600$ s. On the other hand, the proposed virtual inertia controller-based optimal PI controller can address the applicable uncertainties and change the frequency rapidly to its nominal value (50 Hz). Moreover, in this extreme case of uncertainty, the proposed coordination scheme can maintain the dynamic security of the μG and the stable frequency response compared to other comparable methods.

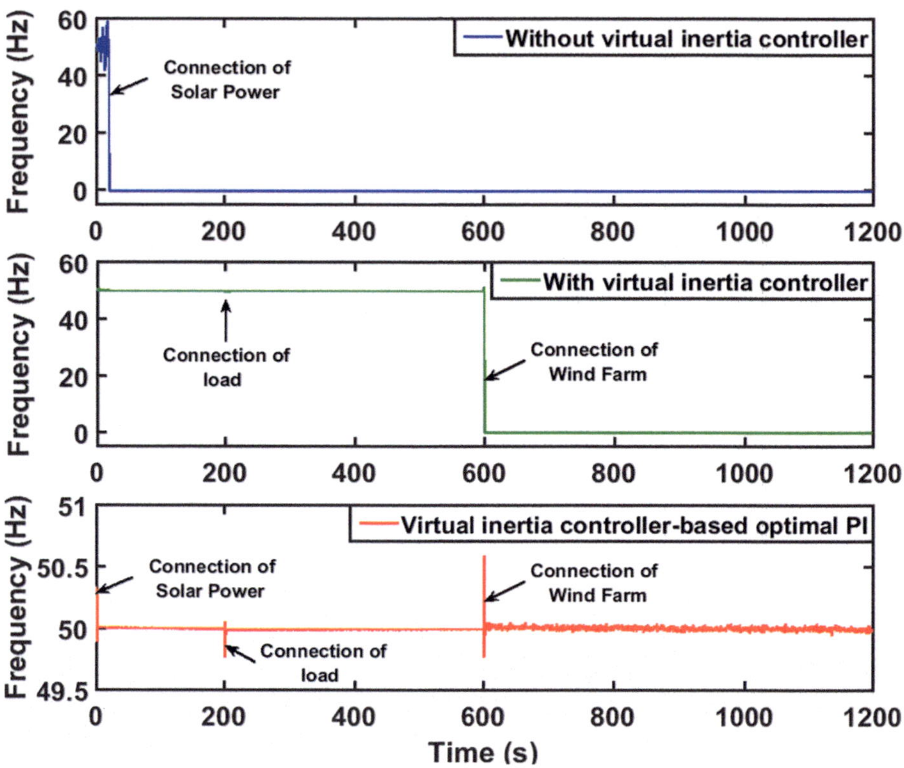

Fig. 4.19 The frequency response of the studied μG for Scenario 3D (i.e., 30% of default system inertia)

Table 4.7 The performance specifications of the studied μG for Scenario D

Scenario D	Virtual inertia controller-based optimal PI controller			With virtual inertia controller			Without virtual inertia controller		
	MUS (pu)	MOS (pu)	T_S (s)	MUS (pu)	MOS (pu)	T_S (s)	MUS (pu)	MOS (pu)	T_S (s)
High system inertia (100%)	0.141	0.331	4	0.290	0.761	20.0	0.330	0.980	24.0
Medium system inertia (50%)	0.195	0.462	4	0.501	1.320	21.0	–	–	–
Low system inertia (30%)	0.220	0.500	5	–	–	–	–	–	–

The performance specifications, (MOS), (MUS), and (T_S) of the studied μG, with different control strategies (i.e., virtual inertia control-based optimal PI controller, virtual inertia controller, and without virtual inertia controller) for Scenario D during the whole period of simulation (20 min) have been compared in Table 4.7.

4.7 Conclusion

Due to the rapid increase in the use of RESs based on the power converter, the overall system inertia in the islanded μGs might be significantly reduced, and thus may threaten the dynamic security of μG. Therefore, this study proposes a novel coordination scheme of frequency control including virtual inertia control and digital OUFR for enhancement of the frequency stability and preservation of the μG dynamic security due to the high penetration level of RESs. The proposed virtual inertia controller based on the optimal PI controller that is optimally designed by using PSO has been applied to emulate inertia characteristic, which contributes to the total inertia of the islanded μG, enhancing the frequency stability and resilience. The simulation results reveal that the proposed optimal PI controller-based virtual inertia controller can effectively regulate the μG frequency and guarantee robust performance compared to with/without the virtual inertia controller, and thus the proposed coordination scheme has achieved effective performance to maintain the dynamic security of the μG. Also, the results confirmed that the digital frequency protection device has superior accuracy, sensitivity, and extensive control.

Chapter 5
A New Trend in Control of Renewable Power Systems Based on Virtual Synchronous Generator

5.1 Introduction

In traditional power systems, the SGs are responsible for providing inertia to the grid through the stored kinetic energy in their rotating mass. Also, they provide damping property for the system that results from the mechanical friction and electrical losses [3]. The intrinsic kinetic energy and damping property of the SGs play an important role in the power system stability. On the other hand, in RPSs, the RESs may replace many SGs; thus the inertia of RPSs decreases due to lack of any rotating mass, which is the main source of inertia [85]. Therefore, with increasing the penetration level of RESs into the power systems, the influence of low system inertia and damping effect on the dynamic system performance and stability increases. Moreover, the intermittent nature of the RESs causes many control problems such as frequency/voltage instability problem, which may limit their high penetration [44]. As a result, RPSs become more susceptible to the disturbances than traditional power systems, and thus face some of the disturbances that threaten their dynamic security such as large frequency/voltage fluctuations, sudden load shedding, forced islanding incidents, and short circuit faults with long clearing times [3].

To overcome such a problem, one of the advanced arrangements is to emulate the behavior of conventional SG in power systems to enhance the system inertia, stability, and flexibility. It is known as a VSG that mimics the activity of the prime movers, thus adding some inertia to the system control loop virtually and accordingly stabilizing the system frequency [2], where it can be established by using ESS together with a power electronic inverter and a proper control mechanism [85]. Therefore, the concept of VSG can provide a basis for maintaining a large share of RESs in today's RPSs without sacrificing system stability and resiliency. Some of the research and studies on VSG applications for RPSs have been conducted in [115–121]. The ref. [115] focused on the modeling of the VSG for the power electronic inverter-based RESs. The frequency stability analysis for μGs in the presence of VSG has been conducted in [117–119]. The influence of the VSG

G. Magdy et al., *Renewable Power Systems Dynamic Security*, Power Systems,
https://doi.org/10.1007/978-3-030-33455-0_5

model on grid stability has been studied in [120] and [121]. On the other side, various control techniques have been implemented based on the virtual inertia control, which is a particular case of VSG execution such as conventional control based on the derivative controller [103], FLC [104, 105], robust controller based on H_infinite technique [44], and MPC [45] for μG frequency stabilization considering high share of the RESs. Based on the aforementioned researches, the dynamic security of RPSs is not achieved yet as they didn't take into consideration the protection devices' role besides the frequency control loops, primary frequency control (i.e., governor action), secondary frequency control (i.e., LFC), and supplementary frequency control (i.e., virtual inertia control-based VSG). Moreover, the power system protection issue is considered as one of the most important issues that help to achieve the dynamic security of RPSs during high penetration of RESs. The dynamic security of RPSs deals with disturbances that result from low system inertia such as large frequency and voltage fluctuations, loss of generation source, forced load shedding, and short circuit faults [3]. Hence, there are two sides for the preservation of the dynamic security of RPSs (i.e., system synchronism), which are frequency stability and protection of the electrical systems as addressed in this research.

Today's power systems are experiencing structural changes induced by the integration of more RESs, thus making it more complex than traditional power systems. Therefore, RPS designers are endeavoring to use digital devices to cope with this complexity, thus improving cost and usability. Also, technological advancements in protective devices have constituted the basis for the emergence of digital protection relays, which enable alternative approaches to standard protection schemes [122]. The digital protection devices have many features compared to traditional schemes such as more accuracy, less sensitivity to temperature, signal storage capability, and high-speed communication [84]. Therefore, the digital protection devices are utilized in many applications of power system operation (i.e., generation, transmission, and distribution systems). Some endeavors have addressed the power system protection issue from the perspective of short circuit fault as in [92], [93], and [123]. The power system protection issue from the perspective of the frequency protection has been addressed in [94–96] and [108]. However, these frequency protection strategies could not damp the frequency deviation and return it to the allowable limit because of energizing the frequency relay once the system frequency becomes beyond the acceptable limit. Such an issue can be overwhelmed by proposing effective coordination of frequency stability and protection as reported in this study.

Based on the above analysis, the research contribution of this study can be summarized in the following aspects:

1. Propose a new frequency control strategy based on VSG, which emulates the characteristics of a real SG, the inertia and damping properties through the concept of virtual rotor, and the frequency control loops (i.e., primary and secondary frequency controls) through the virtual primary and secondary controls, to compensate the reduction in system inertia that results from adding more

RESs (e.g., non-inertia sources), thus stabilizing the system frequency during high penetration of RESs.
2. The proposed virtual inertia control system based on VSG is coordinated with digital frequency protection for improvement of the frequency stability and preservation of the dynamic security of RPSs because of the high share of the RESs.
3. The virtual primary controller (i.e., proportional controller) and virtual secondary controller (i.e., integral controller) is combined to obtain a PI controller, which is optimally designed by using PSO for finding the optimal tuning of the virtual controller parameters, thus enhancing the power system stability and resilience.
4. The uncertainties of RESs and load are taken into consideration in the virtual controller design procedure. Thus, the proposed coordination scheme will ensure an evasion of system instability and collapse.
5. The effectiveness of the proposed coordination scheme is tested and verified through small-scale RPSs (e.g., μG) as well as large-scale RPSs (e.g., real hybrid power system in Egypt).

5.2 VSG Modeling

With increasing the penetration level of RESs into RPSs, the RESs can bring significant impacts to the system inertia, where the RESs exchange power to the RPSs through power electronic devices (i.e., inverters/converters), which are static devices. Therefore, most of the RESs lack a rotating mass, which is the main source of inertia; thus the associated inertia constant is roughly zero [3]. Hence, the influence of low system inertia on the dynamic system performance and stability increases. Furthermore, this low system inertia issue could affect the power system stability and resiliency in the situation of uncertainties, and thus threaten their dynamic security. To beat the issue posed by the inverter-based RESs, the VSG has been presented to emulate the behavior of a real conventional SG, and thus compensates the inertia reduction in RPSs that results from adding more RESs (i.e., non-inertia sources) [117]. Therefore, the concept of VSG is based on reproduction of the dynamic characteristics of a real SG by combining the concept of virtual rotor (i.e., emulating the inertia and damping properties of real SGs) as well as the concept of virtual primary and secondary control (i.e., emulating the primary and secondary frequency control loops of real SGs). On the other hand, the SGs are considered the source of the dynamics in traditional power systems, where the relationship between the electrical and mechanical parts in SGs is expressed by the so-called swing equation:

$$\Delta P_{\mathrm{M}}(s) - \Delta P_{\mathrm{L}}(s) = (2H \cdot s + D_{\mathrm{s}})\Delta f(s) \tag{5.1}$$

where Δf is the frequency deviation, ΔP_{M} is the mechanical power change, ΔP_{L} is the load change, H is the system inertia, and D_{s} is the load damping coefficient. The

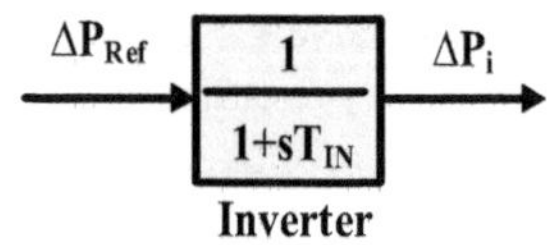

Fig. 5.1 Frequency response model for an inverter

frequency deviation depends mainly on source (turbines) response (i.e., primary control). Once H and/or D_s decrease (increase), the frequency deviation is increased (decreased), where the total inertia constant changes mainly depending on the number of the SGs. Therefore, by increasing the penetration of RESs (i.e., non-inertia sources), the system inertia is decreased. The system frequency control can be divided into three main processes: inertia response process, primary control process, and secondary control process. During the inertia response process (i.e., prior to any controller activation), the SG releases the kinetic energy stored in its rotating mass, when the power unbalance occurs (i.e., difference between turbine power and generator power) [124]. Then, if the frequency deviation exceeds the nominal value, the primary frequency controller is activated immediately. The primary frequency control uses the generator governor to return the frequency to a new steady-state condition within 30 s. Afterward, the secondary frequency control (i.e., LFC) stabilizes the frequency to its nominal value for a time from few seconds to few minutes after the disturbance [8]. Based on the above analysis for the real SG characteristics, the three control processes (i.e., inertia response process, primary control process, and secondary control process) should be emulated for VSG implementation, virtually. Also, the RESs exchange power to RPSs through power inverter, which should be considered for VSG implementation. Therefore, the main components of VSG are the following:

5.2.1 Inverter Model

The inverter is a power electronic device that converts a DC voltage to an AC voltage. It is used for interconnecting RESs to RPSs. The inverter can follow a reference signal by controlling the gates of the insulated gate bipolar transistor using pulse-width modulation method [117]. Therefore, power inverters can follow reference power with a small time constant (T_{IN}), and thus they can be modeled as a first-order transfer function as shown in Fig. 5.1. Hence, the inverter can be considered as an SG with a small time constant and non-inertia response [117, 125].

5.2.2 Virtual Rotor

Based on the given swing equation in Eq. (5.1), which describes the behavior of the rotor dynamics in SGs, the power reference (P_{Ref}) for an inverter can be obtained as [117, 125]

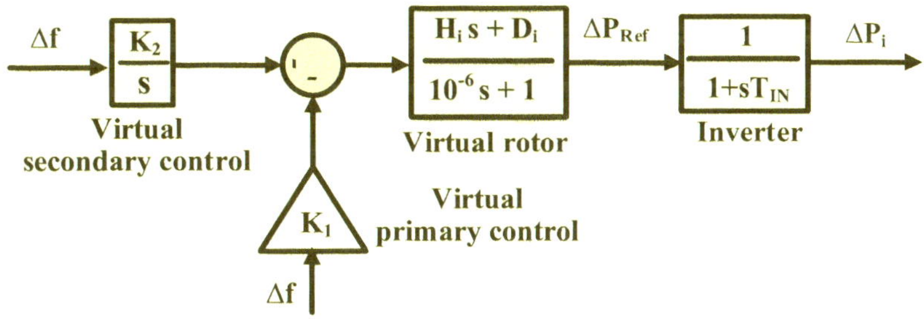

Fig. 5.2 VSG model [125]

$$P_{\mathrm{Ref}} = (H_{\mathrm{i}} \cdot s + D_{\mathrm{i}})\Delta f \tag{5.2}$$

where H_{i} and D_{i} are virtual inertia constant and virtual damping factor, respectively. These parameters are considered as virtual rotor parameters. Hence, the inertia and damping properties of a real SG have been emulated by the virtual rotor model.

5.2.3 Virtual Primary and Secondary Control

As mentioned before, the SG utilizes turbine governor as the primary frequency control to compensate for the deviation of system frequency. The droop control is usually used in power systems to maintain the frequency in allowed range by adjusting the turbine power, which is a decentralized function. Therefore, a similar primary frequency controller (i.e., initial governor response) can be designed for a power inverter. Hence, using a proper controller makes the inverter as an SG, which turbine time constant is very small, and its rotor has zero inertia [117]. On the other hand, the LFC is considered as the major function of AGC systems. It is used in large power systems to stabilize the frequency to its nominal value for a time from few seconds to a few minutes after the disturbance. Therefore, a similar LFC can be designed for a power inverter as a simple integrator controller. Hence, the virtual primary controller (i.e., P controller) and a virtual secondary controller (i.e., I controller) can be combined to obtain a PI controller expressed as

$$\Delta P_{\mathrm{PI}} = \left(K_1 + \frac{K_2}{s}\right)\Delta f \tag{5.3}$$

where K_1 is a virtual primary proportional gain, and K_2 is a virtual secondary integrator gain. Therefore, the model of VSG, which combines virtual rotor, virtual primary frequency control, and virtual secondary frequency control, is shown in Fig. 5.2.

5.3 Virtual Controller Design for the VSG

From the model of VSG, the virtual primary controller and the virtual secondary controller can be combined to obtain a PI controller, which is defined as a virtual controller. The criteria for selecting virtual controller parameters are related to the stability of RPSs during high penetration of RESs. Therefore, this study uses an intelligent searching method (i.e., PSO) to find the optimal tuning of the virtual controller parameters, thus enhancing the RPS stability and resilience. Compared with other methods (i.e., genetic algorithm, artificial neural networks, fuzzy logic, and ant colony) for tuning PI controller, the main advantages of PSO are the ease of use, high convergence rates, minimum storage requirements, and less dependence on the set of initial values, implying the robustness [111]. Moreover, it is computationally inexpensive since its memory and CPU speed requirements are low [126]. Considering these advantages, this study uses the PSO algorithm to tune the virtual controller, thus obtaining the optimum PI controller parameters with the robustness of operations.

5.3.1 Control Strategy and Problem Formulation

In this study, the control strategy of the VSG is based on the virtual controller (i.e., PI controller), which produces the control signal to the VSG for emulating the inertia response into RPSs, thus improving the frequency performance and stability, and avoiding system collapse during contingencies. The virtual controller is composed of proportional gain K_1 and integral gain K_2. Its transfer function is expressed as follows:

$$G_c(s) = K_1 + \frac{K_2}{s} \tag{5.4}$$

The PSO is used to find the optimal design parameters of the virtual controller for minimizing the system frequency deviation, thus enhancing the power system stability and resilience. In this study, the ISE is used as an objective function of the proposed optimization technique and can be formulated as follows:

$$\text{ISE} = \int_0^{t_{\text{sim}}} (\Delta f)^2 \mathrm{d}t \tag{5.5}$$

This is subject to bounds of the virtual controller parameters as [74, 125]

$$\left[K_{1,2}^{\text{Min}} \leq K_{1,2} \leq K_{1,2}^{\text{Max}}\right]$$

where Δf is the frequency deviation of the RPS and t_{sim} is the simulation time to execute one run. The presented PSO algorithm is applied in RPS to obtain the minimum value of the objective function through getting the optimal virtual controller parameters. The details of the PSO algorithm have already been given in Chap. 4.

5.4 Modeling of Digital OUFR

The used model of the digital OUFR has already been described in Chap. 3. Moreover, the principal operation of the OUFR has also been explained in Chap. 3. Hence, the operating sequence of the proposed coordinated scheme of frequency control including virtual inertia control and digital OUFR for preservation of the μG dynamic security is illustrated in Fig. 5.3.

5.5 Test System 1: Microgrid (μG)

5.5.1 System Configuration

An islanded μG is considered as a test system to design and validate the proposed coordination scheme of frequency control including the virtual inertia control based on VSG and digital OUFR. The test system consists of a thermal power plant with 20 MW, wind power with 8 MW, solar power with 4 MW, and domestic loads with 15 MW. The system base is 20 MW. The simplified model of the islanded μG with the proposed coordination scheme is shown in Fig. 5.4. Moreover, this study takes into consideration the important inherent nonlinearity requirements and the physical constraints enjoined by the system dynamics of the generation units, which are GRCs and GDB, to obtain an accurate perception of the studied μG. Therefore, the dynamic model of the islanded μG with the proposed coordination scheme is shown in Fig. 5.5, and the islanded μG parameters are given in Table 5.1.

The dynamic models of wind, solar, and load power are considered in this study. The dynamic models of wind and load power have already been described in Chap. 2. The solar power generation can be represented by an equivalent PV generation plant whose rating is equal to the sum of the ratings of the individual PV generating units. However, the output power of the PV generation system is irregular due to dependence on weather conditions. Hence, the power fluctuations from the PV solar power generation units can be estimated by considering the deviation from the uniform and nonuniform insolation as shown in Fig. 5.6. To obtain an accurate power output profile of a PV solar irradiation model in the RPS, the original random output fluctuation that can be modeled by the white noise block in MATLAB program is multiplied by the standard deviation [127]. The solar power

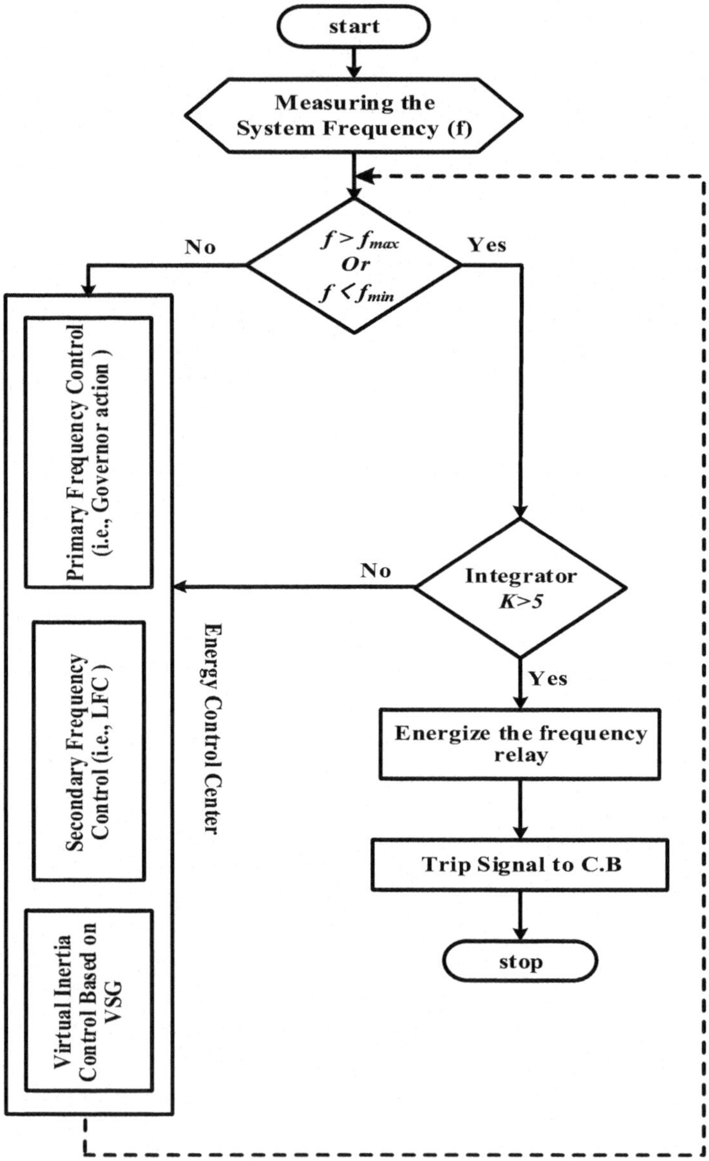

Fig. 5.3 The diagram for implementation of the proposed coordinated scheme [125]

deviation is simulated close to an actual solar power change by the following function:

$$\Delta P_{\text{Solar}} = 0.6\sqrt{P_{\text{Solar}}} \tag{5.6}$$

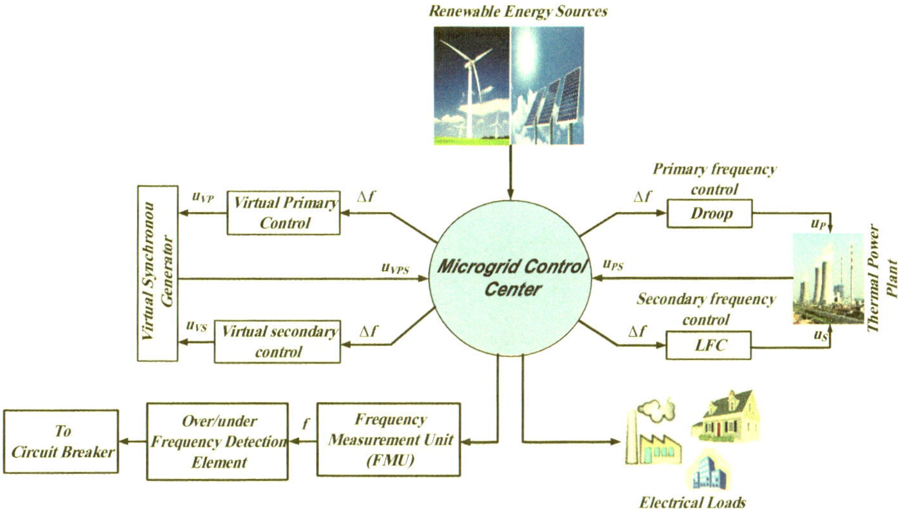

Fig. 5.4 A simplified model of the islanded μG with the proposed coordination scheme [125]

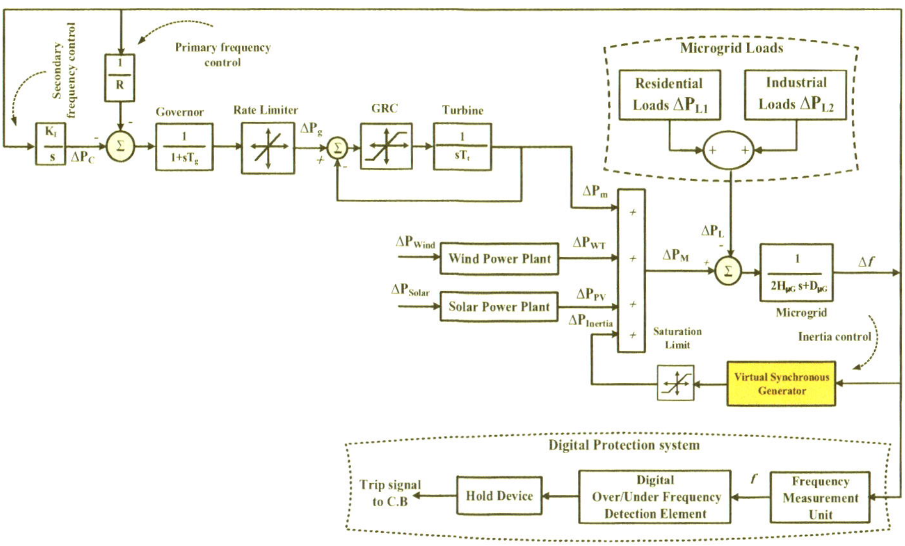

Fig. 5.5 A dynamic model of the studied islanded μG with the proposed coordination scheme [125]

5.5.2 *Implementation of PSO for VSG Design*

In this study, the main objective of the PSO algorithm is to minimize the objective function as given in Eq. (5.5) through finding the optimal parameters of the virtual

Table 5.1 Dynamic parameters of the islanded μG

Parameter	Value	Parameter	Value
$D_{\mu G}$	0.015	K_I	0.05
$H_{\mu G}$	0.083	V_U	0.3
T_g	0.1	V_L	−0.3
T_t	0.4	GRC	20%
R	2.4	f	50

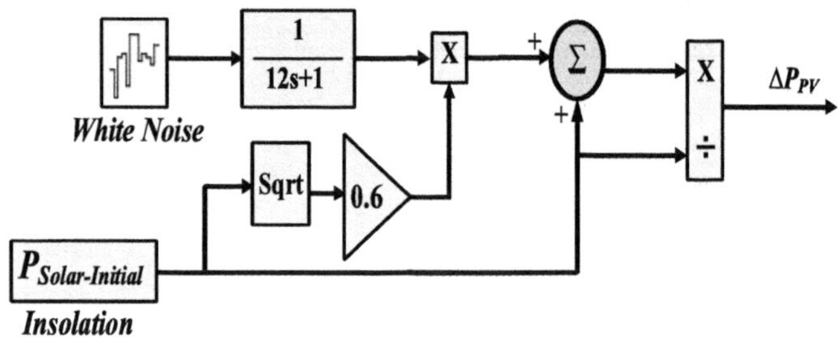

Fig. 5.6 The model of PV solar power using MATLAB/Simulink

Table 5.2 Specification of the PSO technique

Parameter	Value
Size of swarm (i.e., no. of birds)	50
Number of iterations (n)	50
Inertia weight factor (w)	0.95
Acceleration constant 1 (C_1)	0.12
Acceleration constant 2 (C_2)	2

Table 5.3 Parameters of VSG for the studied μG

Parameter	Value	Parameter	Value
H_i	0.9	D_i	10.4
K_1	3.4903	K_2	22.0968
T_{IN}	0.04		

controller for the VSG design. The code of the PSO algorithm is performed using the MATLAB program. The performance of the PSO algorithm in searching the virtual controller parameters of the VSG in the studied μG has been validated by using the characteristics of the PSO as given in Table 5.2. These optimal characteristics are selected by the trial-and-error method, which is the most commonly used with optimization algorithms [45]. Therefore, the parameters of the VSG under the system operation condition are given in Table 5.3. Hence, the optimal virtual controller parameters produce the optimal control signal to the VSG for emulating the inertia response into the μG during high RES penetration.

5.5.3 *Simulation Results and Discussion*

The model of the studied μG with all the details of components as in Fig. 5.5 and Table 5.1 is built using MATLAB/Simulink model to validate the effectiveness and robustness of the proposed coordination scheme. Therefore, the performance of the proposed coordinated scheme of frequency control including the virtual inertia control based on VSG and the digital frequency protection is compared with the high-level operations of primary and secondary frequency controls, known as the conventional control for the studied μG under the nature variety of the RESs, random load variation, and system inertia variations (i.e., system uncertainties). The code of the PSO algorithm as an m-file is interfaced with the model of the islanded μG to execute the optimization process. Maintain the dynamic security of the islanded μG with the proposed coordination scheme investigated under different operating conditions through the following scenarios:

5.5.3.1 Performance Assessment of the Studied μG Without the RESs

Scenario 1: Sudden Load Change

In this scenario, the performance of the islanded μG with the proposed coordination scheme is tested and evaluated by using 2.0 MW step load change ($\Delta P_L = 0.1$ pu) at time $t = 300$ s. This phenomenon can occur in the event of forced outage of generation unit or sudden switch off of a massive load. Moreover, the studied μG with the proposed coordination scheme is tested and evaluated in the circumstances of high system inertia (i.e., 100% of its nominal value), medium system inertia (i.e., 50 of its nominal value), and low system inertia (i.e., 30% of its nominal value). Figure 5.7 shows the frequency response of the studied μG in the circumstances of high, medium, and low system inertia, respectively. In case of either high or medium system inertia, the studied μG with the proposed coordination scheme of frequency control including virtual inertia control based on VSG and digital OUFR is significantly more stable and faster, compared to that with conventional control. The frequency changes of the studied μG with conventional control are ±0.34 Hz and ±0.51 Hz in the circumstances of high and medium system inertia, respectively, while the studied μG with the proposed coordination scheme gives a very little frequency change of ±0.001 Hz in that cases during the connection of load change at $t = 300$ s. On the other hand, in case of low system inertia (i.e., 30% of its nominal value), the virtual inertia control based on VSG significantly enhances frequency performance and decreases the magnitude of system transients. However, the conventional control cannot endure the variation of the μG frequency due to connecting of the load change at $t = 300$ s; thus the digital OUFR energizes and sends a trip signal to the generator circuit breaker. This happened due to the frequency response of the studied μG with conventional control fluctuation beyond the allowable frequency limits when a step load change is connected at $t = 300$ s as well as the

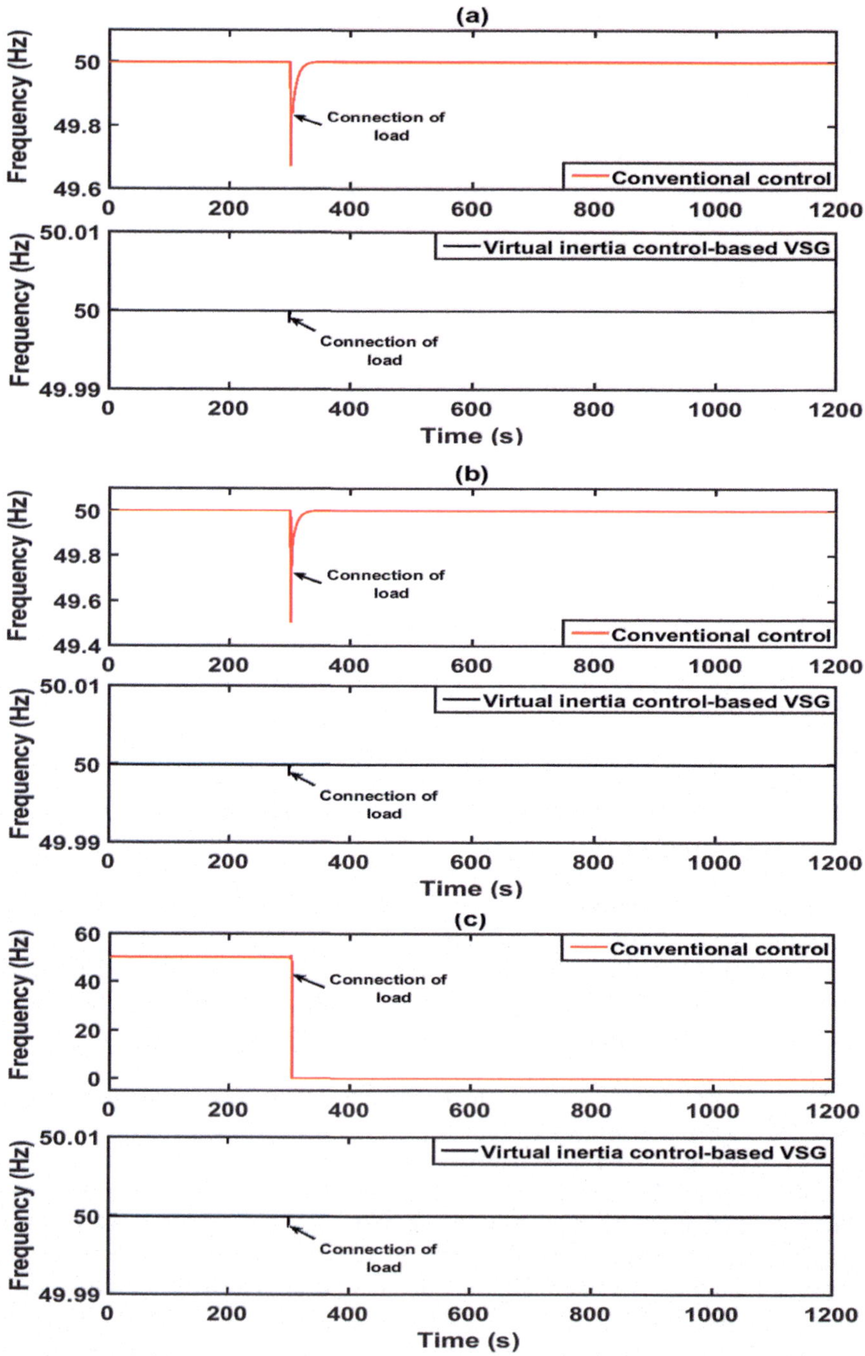

Fig. 5.7 The frequency response of the studied μG for scenario 1. (**a**) High system inertia, (**b**) medium system inertia, and (**c**) low system inertia

integrator output magnitude is larger than the integrator set time ($K = 5$ s). Hence, the preservation of the studied μG dynamic security has been achieved by using the proposed coordination scheme.

Scenario 2: Multiple Disturbances in Load Demand

In this scenario, the performance of the islanded μG with the proposed coordination scheme is investigated by implementing industrial load and residential load as shown in Fig. 5.8. Moreover, the studied μG is tested in the circumstances of high, medium, and low system inertia to represent the significant effect of actual μG operations. Therefore, this studied scenario is implemented under the assumed various operating states of random load variants as shown in Table 5.4. Figure 5.9 shows the frequency response of the studied μG under this studied scenario as well as in the circumstances of high, medium, and low system inertia, respectively. In case of high system inertia (i.e., 100% of its nominal value), the frequency change of the studied μG with the conventional control (i.e., only primary and secondary frequency controls) is ±0.75 Hz, while the studied μG with the dynamic contribution of virtual inertia control based on VSG gives a very little frequency change of ±0.002 Hz during the connection of the industrial load (i.e., high-fluctuated random load variation) at $t = 300$ s. Therefore, the virtual inertia control loop based on VSG can readjust the frequency to its normal value by adding some inertia to the system control loop virtually and accordingly stabilizing the system frequency. On the other hand, in case of either medium or low system inertia, the lack of system inertia significantly affects the stability of the studied islanded μG with conventional control. The conventional control cannot withstand the variation of the μG frequency, and thus the digital OUFR energizes and sends a trip signal to the generator circuit breaker in that case, during connecting of the industrial load at $t = 300$ s under the situation of medium system inertia (i.e., 50% of its nominal value) as shown in Fig. 5.9a, and during connecting of the residential load at $t = 0$ s under the situation of low system inertia (i.e., 30% of its nominal value) as shown in Fig. 5.9b. The virtual inertia control based on VSG significantly enhances frequency performance and decreases the magnitude of system transients. Therefore, the studied μG with the proposed coordination scheme does not need any action from the digital OUFR. Hence, the proposed coordination effectiveness was approved to maintain the μG dynamic security.

5.5.3.2 Performance Assessment of the Studied μG Including the RES Uncertainty

Scenario 3: Sudden Load Change

The main target of this scenario is to evaluate the performance of the studied μG with the proposed coordination scheme under variation in system inertia (i.e., system

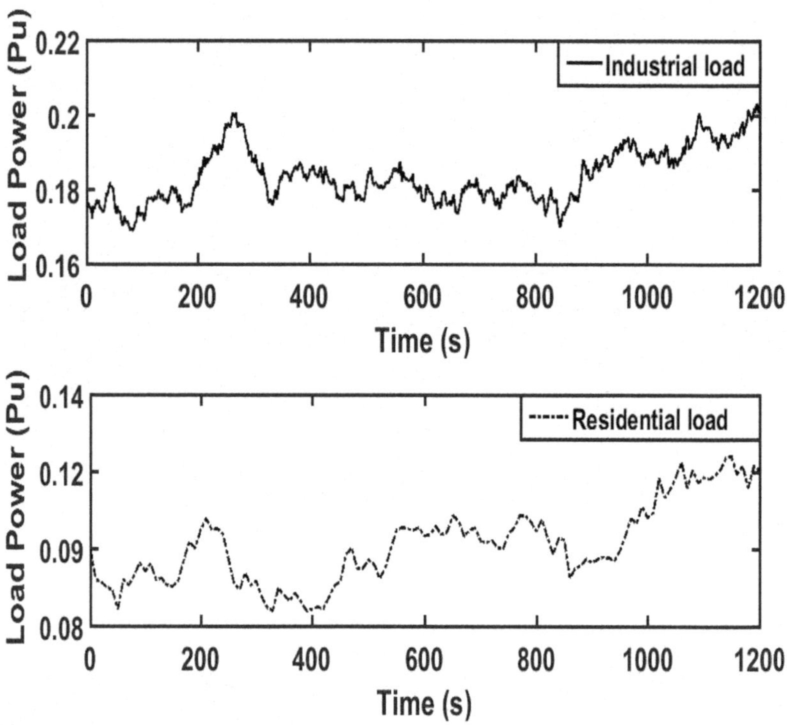

Fig. 5.8 The random load deviation of cases studied

Table 5.4 Multiple operating conditions of the studied μG for scenario 2

Disturbance source	Starting time (s)	Stopping time (s)	Size (pu)
Industrial load	300 s	–	0.21
Residential load	Initial	800 s	0.12

uncertainties) and nature of the RESs, which are defined as the important properties of the actual μG. Therefore, the studied μG is tested by implementing high-fluctuated wind power and low-fluctuated solar power as shown in Fig. 5.10, as well as is implemented under the assumed various operating states in Table 5.5. Moreover, a 10% step load change is implemented to the studied μG at time $t = 300$ s.

Figure 5.11 shows the frequency response of the studied μG considering high penetration of RESs in the circumstances of high, medium, and low system inertia, respectively. From the simulation results, in the situation of high system inertia (i.e., 100% of its nominal value), the frequency change of the μG with the conventional control (i.e., only primary and secondary frequency controls) is ±0.78 Hz, while the studied μG with the virtual inertia control based on VSG gives a very little frequency change of ±0.04 Hz when the wind farm is connected at $t = 600$ s. Therefore, the studied μG considering high penetration of RESs with the proposed coordination scheme is significantly more stable and faster, compared to that with conventional

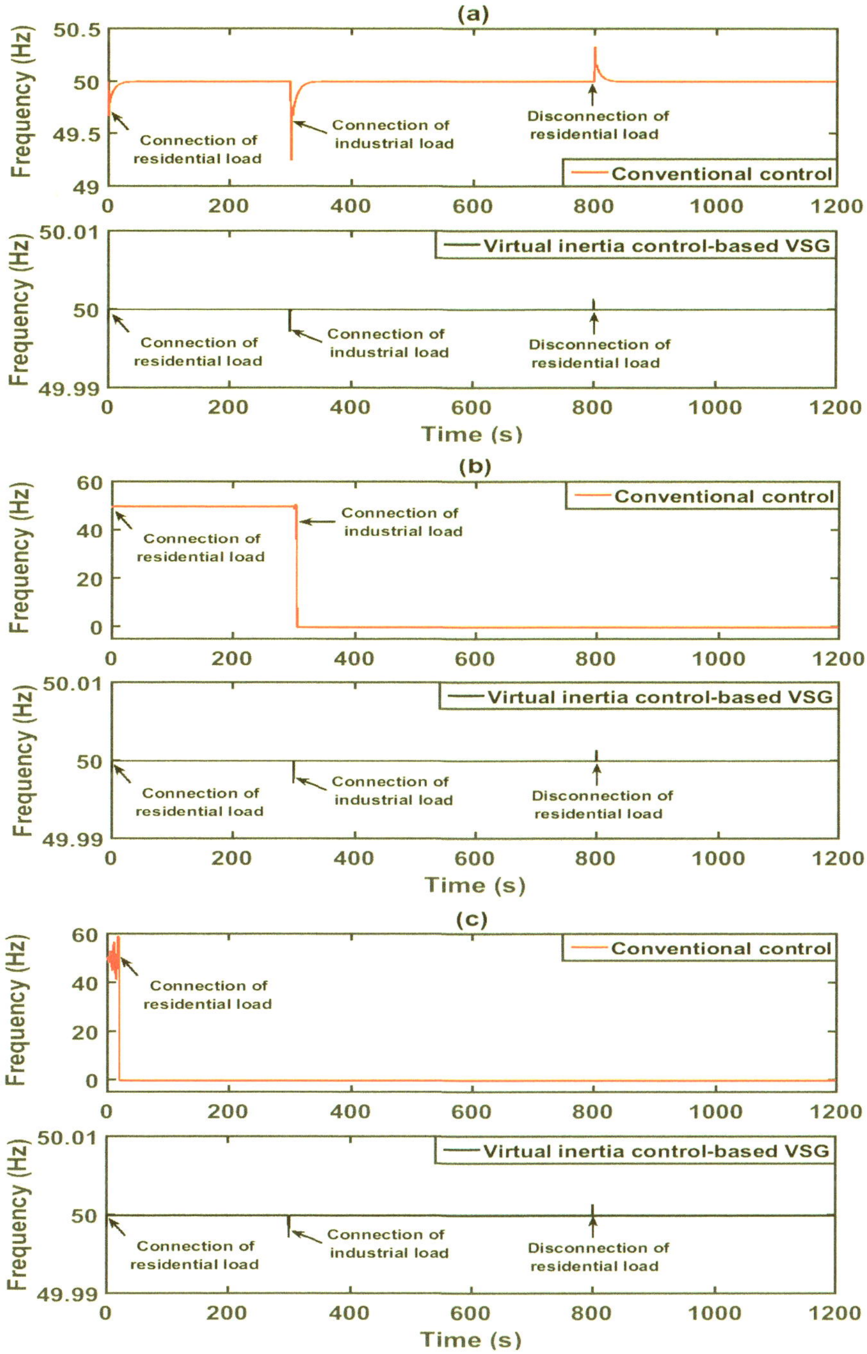

Fig. 5.9 The frequency response of the studied μG for scenario 2. (**a**) High system inertia, (**b**) medium system inertia, and (**c**) low system inertia

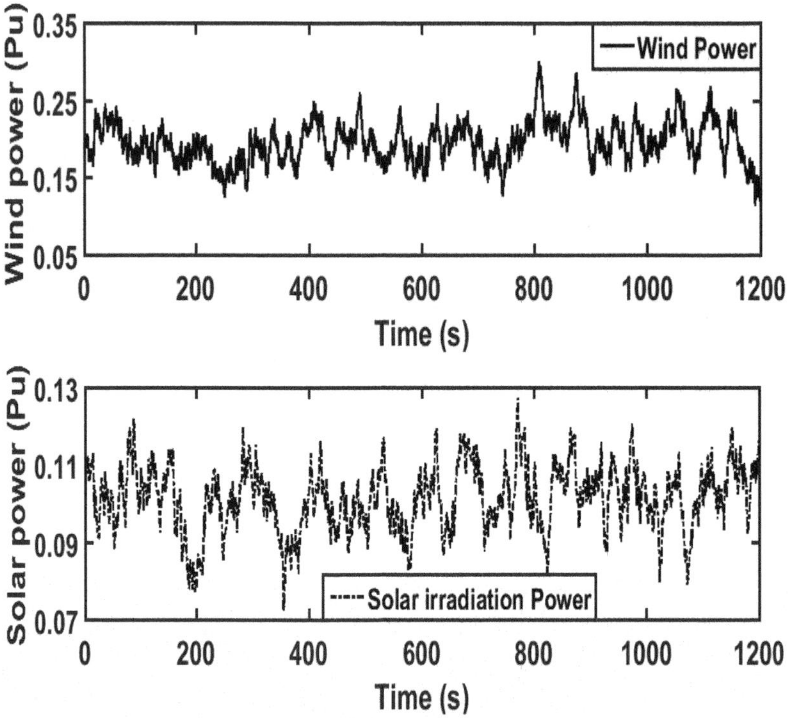

Fig. 5.10 Power variation patterns of wind and solar irradiation power

Table 5.5 Multiple operating conditions of the islanded μG for scenario 3

Disturbance source	Starting time (s)	Stopping time (s)	Size (pu)
Wind farm	600 s	–	0.30
Solar power plant	Initial	–	0.12

control as shown in Fig. 5.11a. On the other hand, in case of either medium or low system inertia, the lack of system inertia significantly affects the stability of the studied μG with conventional control. Conventional control cannot restore the system frequency to the nominal value. Thus, the digital OUFR energizes and sends a trip signal to the generator circuit breaker in that case, during connecting of a wind farm at $t = 600$ under the situation of medium system inertia (i.e., 50% of its nominal value) as shown in Fig. 5.11b, and when connecting the solar power at $t = 0$ s under the situation of low system inertia (i.e., 30% of its nominal value) as shown in Fig. 5.11c. The proposed virtual inertia control based on VSG can properly maintain the change within the acceptable range of ±0.042 Hz and ±0.048 Hz in the circumstances of medium and low system inertia, respectively. Hence, the proposed coordinated scheme of frequency control including the virtual inertia control based on VSG and the digital OUFR can maintain the dynamic security of the studied μG considering high penetration of RESs and the stable frequency response compared to the conventional control.

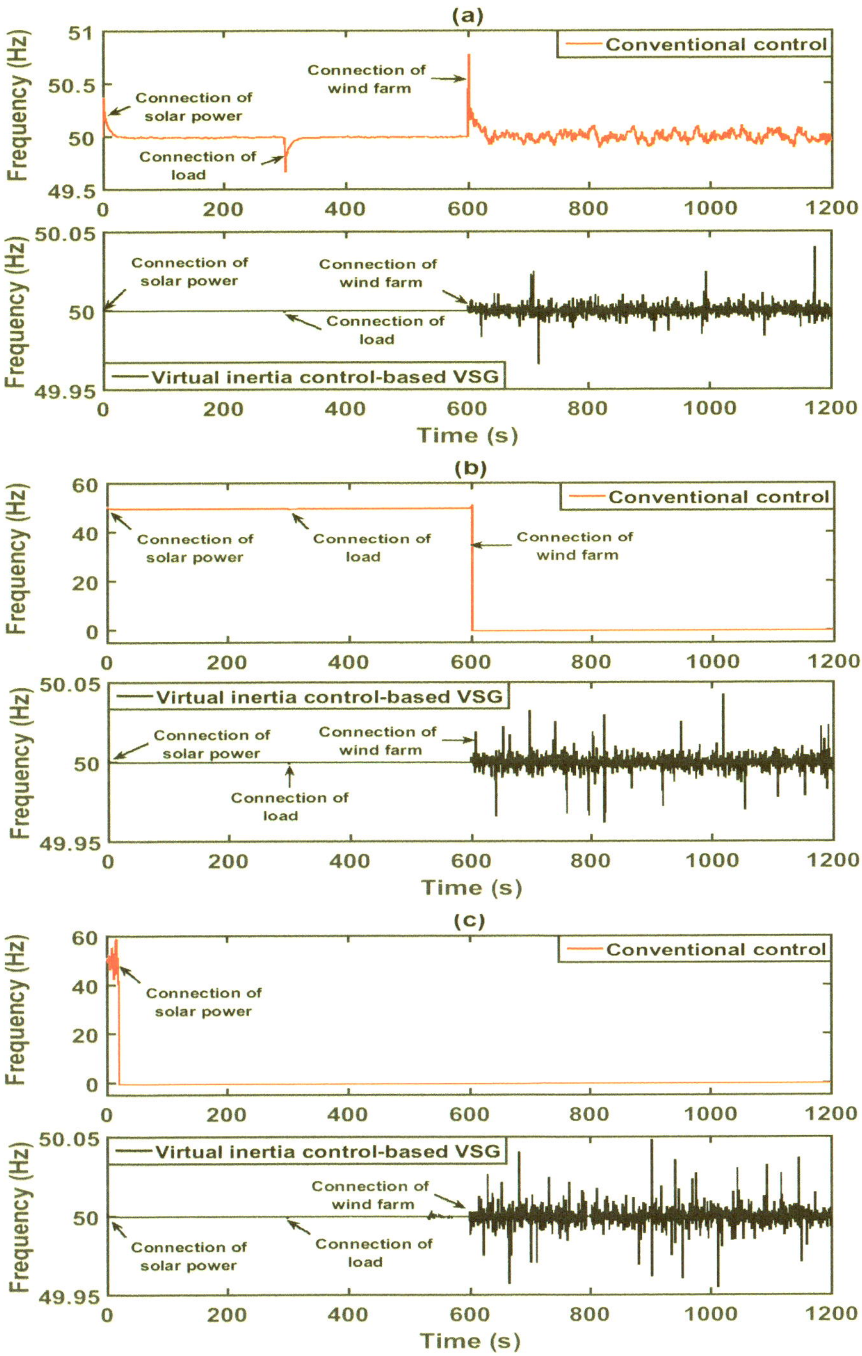

Fig. 5.11 The frequency response of the studied μG for scenario 3. (**a**) High system inertia, (**b**) medium system inertia, and (**c**) low system inertia

Scenario 4: Multiple Disturbances in Load Demand

In this scenario, the studied μG with the proposed coordination scheme is investigated in the presence of high-fluctuated wind power and low-fluctuated solar power as shown in Fig. 5.10, and also high-fluctuated random load variation (i.e., industrial loads) and low-fluctuated random load variation (i.e., residential load) as shown in Fig. 5.8. Moreover, the studied μG is tested in the circumstances of high, medium, and low system inertia to represent the significant effect of actual μG operations. Therefore, this scenario is implemented under the assumed various operating states of random load and RES variants given in Tables 5.4 and 5.5.

Figure 5.12 shows the frequency response of the studied μG considering high penetration of RESs and random load in the circumstances of high, medium, and low system inertia, respectively. In the case of high system inertia (i.e., 100% of its nominal value), the studied μG frequency fluctuates within the allowable frequency limits. Therefore, in this case, the role of the frequency control loops emerges without any action from the digital OUFR. From Fig. 5.12a, it is obvious that the frequency change of the studied μG with the conventional control is ±0.78 Hz when the wind farm (i.e., high-fluctuated wind power) is connected at $t = 600$ s. The proposed virtual inertia control based on VSG enhances the frequency performance (i.e., the frequency change of about ±0.038 Hz) and reduces the magnitude of system transients compared to the studied μG with the conventional control. On the other hand, in case of either medium or low system inertia, the frequency response of the studied μG with the conventional control fluctuates severely due to critical inertia reduction. Also, the integrator output of the digital OUFR overtakes the integrator set time *K*. Thus, the digital OUFR energizes and sends a trip signal to the generator circuit breaker in that case, during connecting of the industrial load at $t = 300$ s under the situation of medium system inertia (i.e., 50% of its nominal value) as shown in Fig. 5.12b, and under the situation of low system inertia (i.e., 30% of its nominal value) as shown in Fig. 5.12c. Compared to the studied μG with the conventional control, the proposed virtual inertia control based on VSG can address the applicable uncertainties and the change of the frequency rapidly to its nominal value (50 Hz). Furthermore, in this extreme case of uncertainty, the proposed coordination scheme can maintain the dynamic security of the μG.

5.6 Test System 2: Real Hybrid Power System in Egypt

5.6.1 System Configuration

The real hybrid power system in Egypt (i.e., large-scale power system) is considered as a test system to validate the effectiveness of the proposed coordinated scheme of frequency control including the virtual inertia control based on VSG and digital OUFR. The EPS includes both conventional generation sources (i.e., thermal, gas, and hydraulic power plants) with inherent nonlinearities and RESs (i.e., wind and solar energy). The conventional power plants can be classified into three categories:

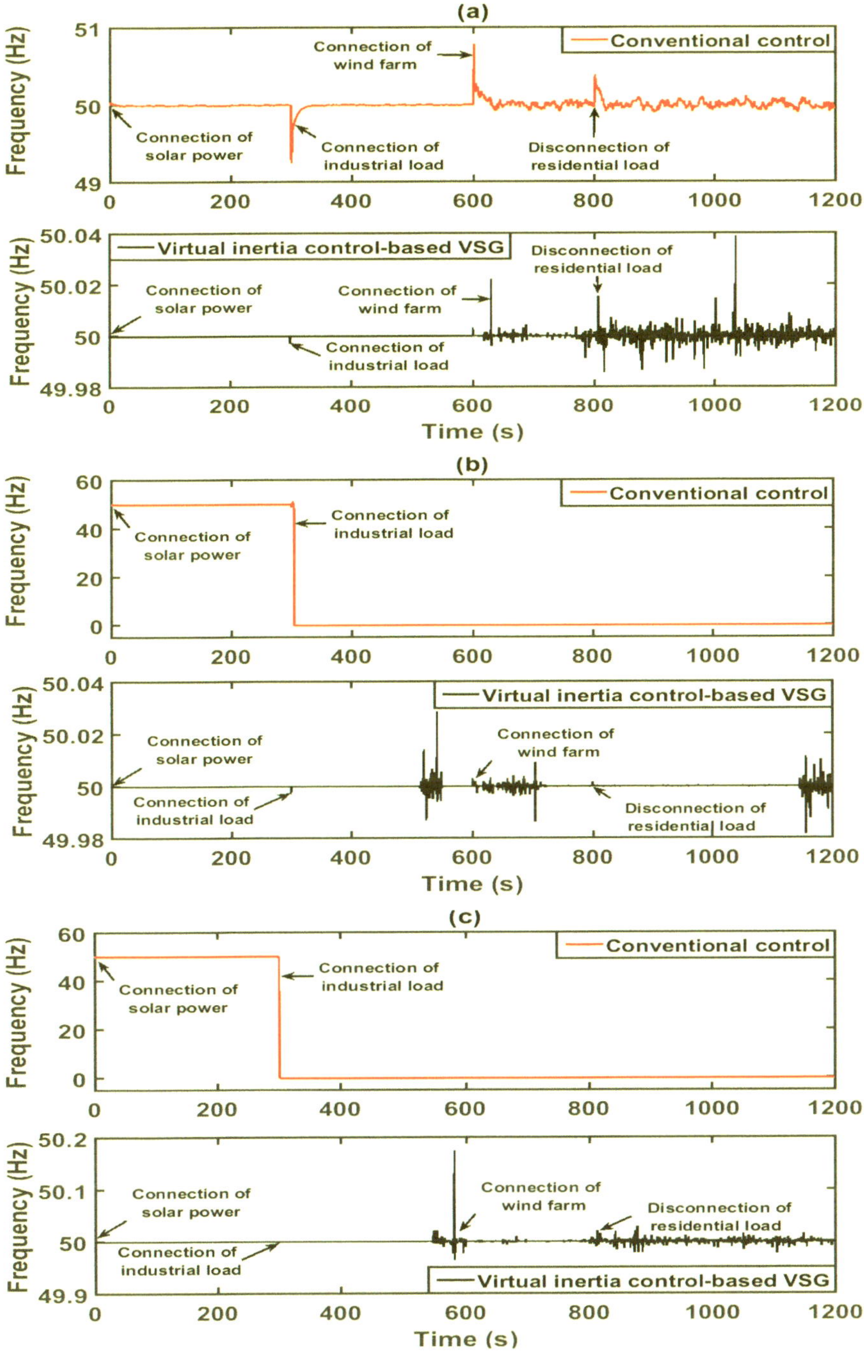

Fig. 5.12 The frequency response of the studied μG for scenario 4. (**a**) High system inertia, (**b**) medium system inertia, and (**c**) low system inertia

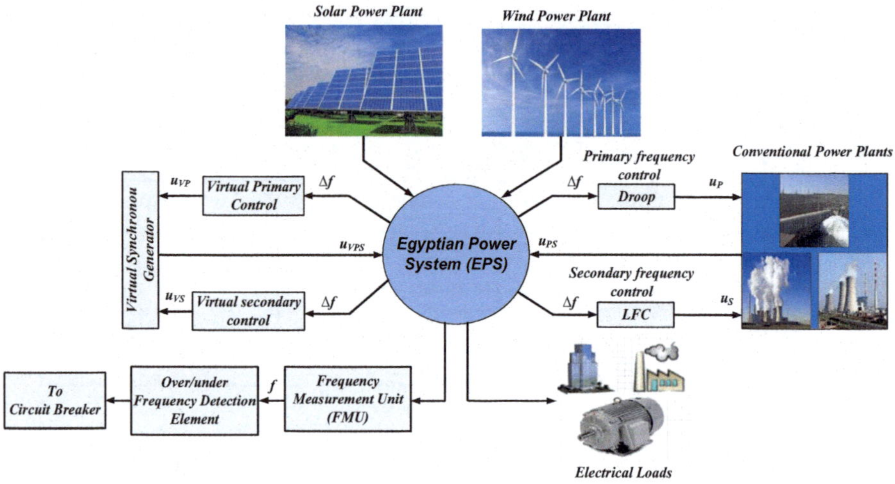

Fig. 5.13 A simplified model of the RPS in Egypt with the proposed coordination scheme [125]

(a) non-reheat power plants represented by gas-turbine power plants and a few numbers of steam power plants; (b) reheat power plants mainly represented by thermal power plants or combined cycle power plants; and (c) hydraulic power plants such as High Dam in Aswan city. According to the last report of the Egyptian Electricity Holding Company (EEHC) in 2017, the total generation capacity and peak loads are 45,008 MW and 29,400 MW, respectively [128]. Moreover, 5% of the installed capacity is extracted from the RESs but the Ministry of Electricity and Renewable Energy of Egypt plans to increase the electric energy from RESs to cover 42% of the electric energy demand by the year 2030 [128]. Therefore, this study considers the EPS with consideration of the effect of the high penetration level of RESs (i.e., future planning of the EPS) as a test system to validate the effectiveness of the proposed coordination scheme. The simplified model of the RPS in Egypt with the proposed coordination scheme is shown in Fig. 5.13.

On the other side, the important inherent nonlinearity requirements and the physical constraints enjoined by the system dynamics of the generation units such as GRC and GDB are taken into consideration to obtain an accurate perception for the EPS. The GRC limits the generation rate of output power, which is given as 20% pu MW/min and 10% pu MW/min for non-reheat and reheat turbines, respectively. However, the actual GRC of the hydraulic power plant is about 50% pu MW/min, which is higher than the generation rate corresponding to any practical disturbance, and hence it will be neglected [129]. Moreover, the previously described dynamic models of wind, solar energy, and load power are considered in this study. Therefore, the dynamic model of the EPS with the proposed coordination scheme is shown in Fig. 5.14, and the system parameters are given in Table 5.6.

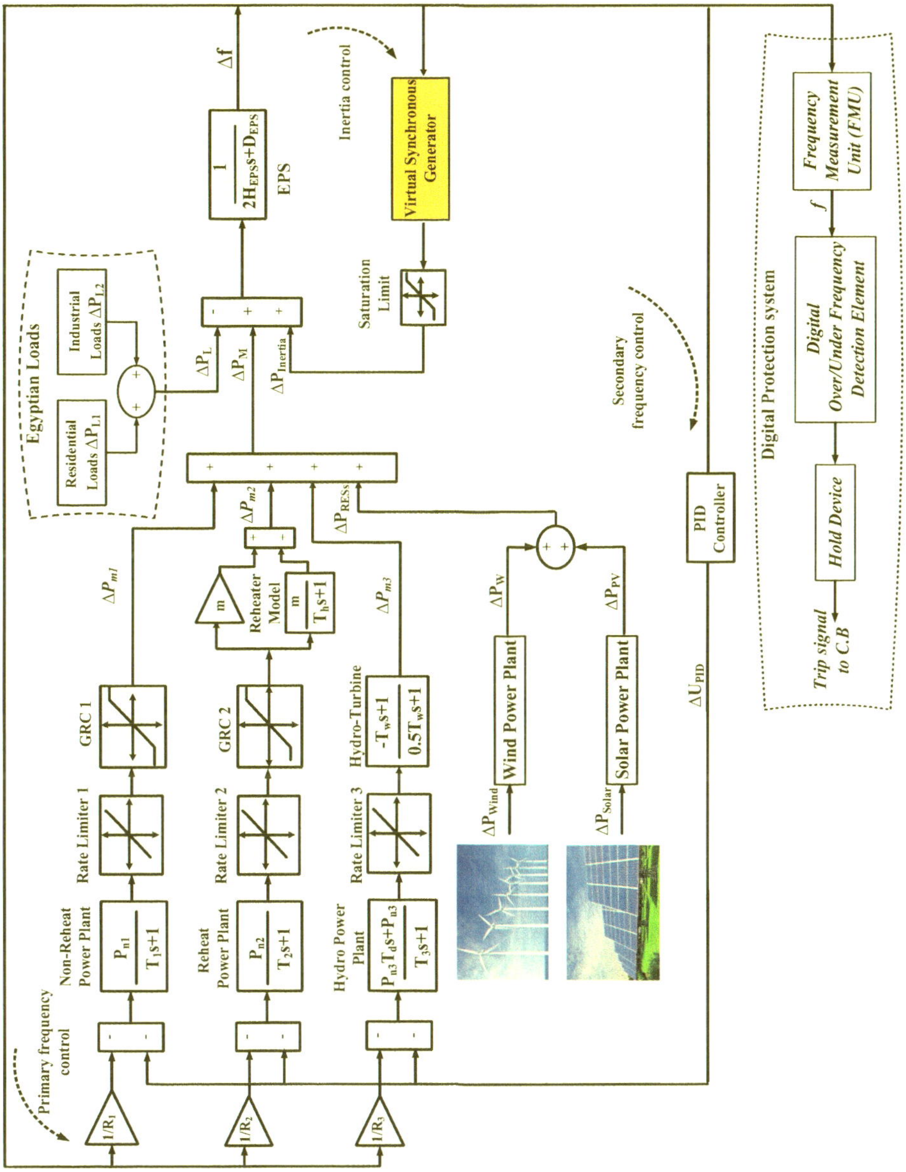

Fig. 5.14 A dynamic model of the EPS with the proposed coordination scheme [125]

5.6.2 Implementation of PSO for VSG Design

In this section, the PSO algorithm has been adapted to find the optimal parameters of the virtual controller of the VSG using the model of the EPS through minimization of

Table 5.6 Dynamic parameters of the EPS

Parameter	Value	Parameter	Value	Parameter	Value
H_{EPS}	5.7096	T_w	1.0	P_{n3}	0.1364
D_{EPS}	0.028	m	0.5	K_p	71.2532
T_1	0.4	R_1	2.5	K_i	5.9055
T_2	0.4	R_2	2.5	K_d	6.10758
T_3	90	R_3	1.0	f	50
T_d	5	P_{n1}	0.2529		
T_h	6	P_{n2}	0.6107		

the objective function as given in Eq. (5.5). The PSO algorithm had superior features such as easy implementation and stable convergence characteristics and it can generate a high-quality solution within shorter computation time. However, the selection of PSO parameters decides to a great extent the ability of global minimization. The performance of the PSO in searching the virtual controller parameters of the VSG in the EPS has been validated by using the characteristics of the PSO as given in Table 5.2. Therefore, the parameters of the VSG under the system operation condition are given in Table 5.7. Hence, the optimal virtual controller parameters produce the optimal control signal to the VSG for emulating the inertia response into the EPS control loop, thus stabilizing the system frequency during high penetration of RESs.

5.6.3 *Simulation Results and Discussion*

The simulation results of the EPS considering high penetration level of RESs are carried out using MATLAB/Simulink® software to validate the effectiveness and robustness of the proposed coordination scheme. Moreover, the performance of the proposed coordinated scheme of frequency control loops including the virtual inertia control based on VSG and the digital OUFR is compared with the conventional control (i.e., primary and secondary frequency control loops) under the nature variety of the RESs, random load variation, and system inertia variations (i.e., system uncertainties). Therefore, maintaining of the dynamic security of the EPS with the proposed coordination scheme is investigated under different operating conditions through the following scenarios:

5.6.3.1 Performance Assessment of the EPS Without the RESs

Scenario 1: Sudden Load Change

The studied large-scale RPS (i.e., the EPS) with the proposed coordination scheme is tested and evaluated in the circumstances of high system inertia (i.e., 100% of its

Table 5.7 Parameters of VSG for the EPS

Parameter	Value	Parameter	Value
H_i	0.9	D_i	10.4
K_1	7.48710	K_2	1.17135
T_{IN}	0.04		

nominal value), medium system inertia (i.e., 50% of its nominal value), and low system inertia (i.e., 30% of its nominal value). Moreover, a 10% step load change is implemented to the EPS at time $t = 300$ s to simulate the forced outage of the generation unit or sudden switch off of a massive load. Figure 5.15 shows the frequency response of the studied EPS in the circumstances of high, medium, and low system inertia, respectively. From the simulation results, in the situation of high system inertia, the virtual inertia control based on VSG significantly enhances frequency performance and decreases the magnitude of system transients, compared to the studied EPS with the conventional control as shown in Fig. 5.15a. On the other hand, in case of either medium or low system inertia, the lack of system inertia significantly affects the stability of the studied EPS with conventional control. The conventional control cannot endure the variation of system frequency due to connecting of the load change at $t = 300$ s; thus the digital OUFR energizes and sends a trip signal to the generator circuit breaker as shown in Fig. 5.15b, c. The studied EPS with the proposed coordination scheme based on the virtual inertia control-based VSG gives a very little frequency change in that case during the connection of load change at $t = 300$ s. Hence, the proposed coordinated scheme of frequency control including the virtual inertia control based on VSG and the digital OUFR can maintain the dynamic security of the EPS and the stable frequency response compared to the conventional control.

Scenario 2: Multiple Disturbances in Load Demand

The studied EPS with the proposed coordination scheme is tested in the circumstances of high, medium, and low system inertia to represent the significant effect of actual RPS operations. Moreover, the performance of the studied EPS with the proposed coordination scheme is investigated by implementing industrial and residential loads as shown in Fig. 5.8. Therefore, this scenario is implemented under the assumed various operating states of random load variants given in Table 5.4. The frequency response of the studied EPS by considering the random loads in the circumstances of high, medium, and low system inertia, respectively, is shown in Fig. 5.16. In the situation of high system inertia (i.e., 100% of its nominal value), the system frequency fluctuates within the allowable frequency limits. Moreover, the proposed virtual inertia control based on VSG enhances the frequency performance and reduces the magnitude of system transients compared to the studied EPS with the conventional control as shown in Fig. 5.16a. On the other hand, in case of either medium or low system inertia, the conventional control (i.e., only primary and secondary frequency controls) cannot withstand the variation of the system

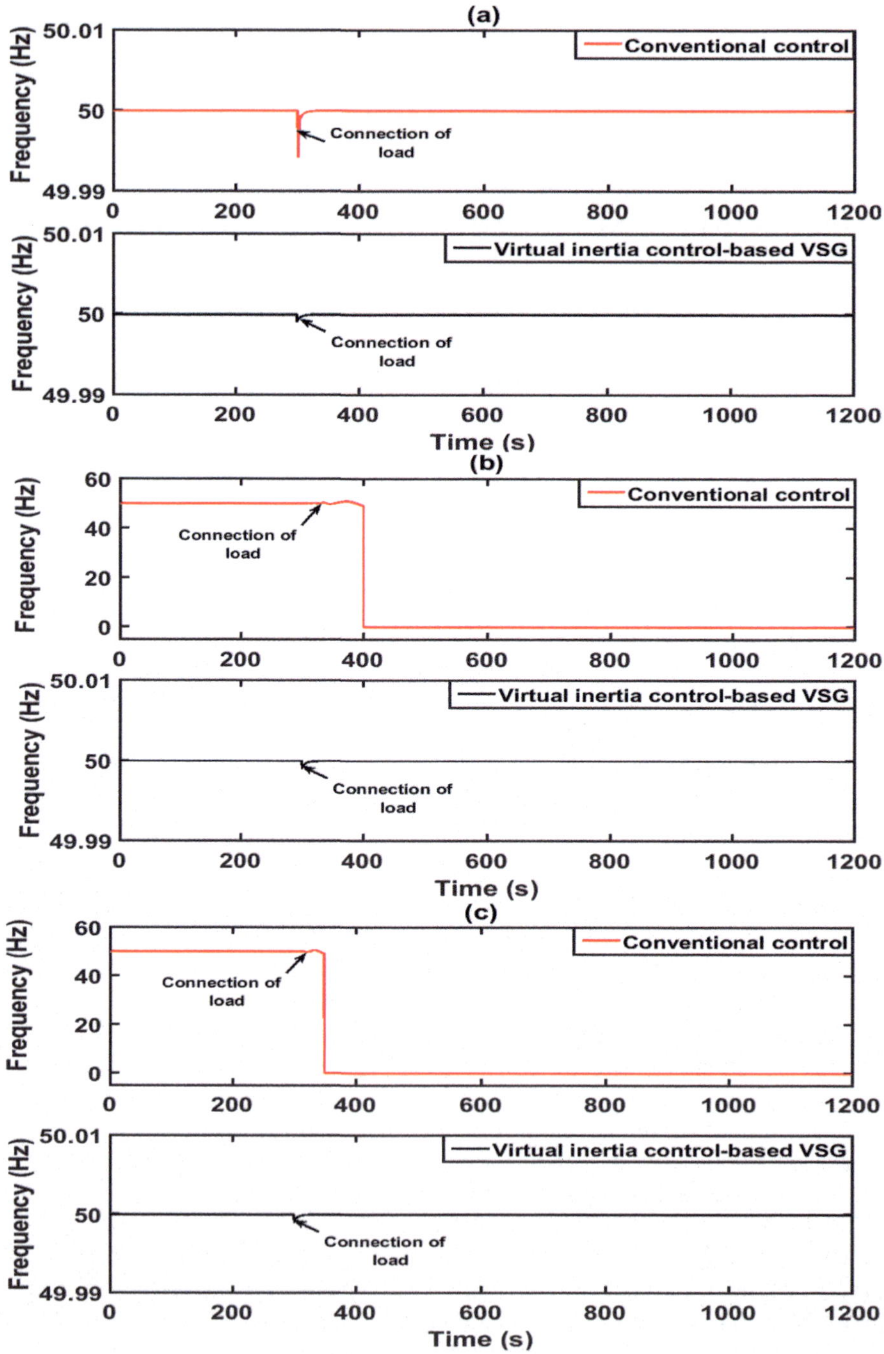

Fig. 5.15 The frequency response of the studied EPS for scenario 1. (**a**) High system inertia, (**b**) medium system inertia, and (**c**) low system inertia

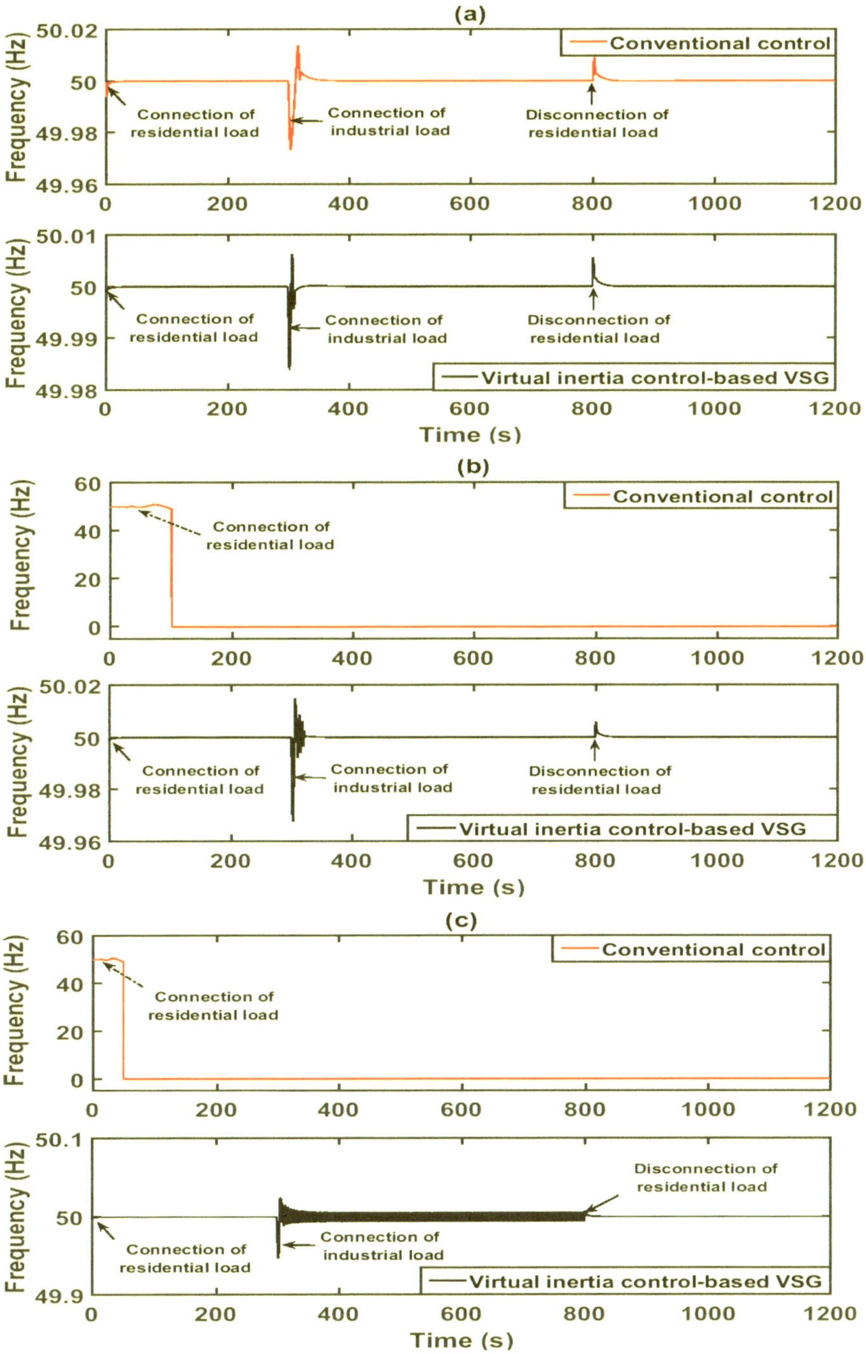

Fig. 5.16 The frequency response of the studied EPS for scenario 2. (**a**) High system inertia, (**b**) medium system inertia, and (**c**) low system inertia

frequency when the residential load is connected at $t = 0$ s as shown in Fig. 5.16b, c. Therefore, the digital OUFR energizes and sends a trip signal to the generator circuit breaker at that time. The proposed virtual inertia control based on VSG can properly maintain the frequency change within the acceptable range and decreases the magnitude of system transients. The frequency changes of the studied EPS with the proposed coordination scheme are ± 0.03 Hz and ± 0.052 Hz in the circumstances of medium and low system inertia, respectively, during connecting of the high-fluctuated random load variation (i.e., industrial loads) at $t = 300$ s.

5.6.3.2 Performance Assessment of the EPS Including the RES Uncertainty

Scenario 3: Sudden Load Change

In this scenario, the studied EPS with the proposed coordination scheme is investigated in the presence of high-fluctuated wind power and low-fluctuated solar power as shown in Fig. 5.10, considering the assumed various operating states in Table 5.5. Moreover, a 10% step load change is implemented to the studied EPS at time $t = 300$ s. Therefore, the performance of the studied EPS with the proposed coordination scheme is tested and evaluated under variation in system inertia: high, medium, and low system inertia. Figure 5.17 shows the frequency response of the studied EPS considering high penetration of RESs in the circumstances of high, medium, and low system inertia, respectively. In the situation of high system inertia (i.e., 100% of its nominal value), the frequency change of the studied EPS with the conventional control is ± 0.0215 Hz, while the studied EPS with the proposed coordination scheme based on the virtual inertia control-based VSG gives a very little frequency change of ± 0.007 Hz when the wind farm is connected at $t = 600$ s. Therefore, the studied EPS considering high penetration of RESs with the proposed coordination scheme is significantly more stable and faster, compared to that with conventional control as shown in Fig. 5.17a. On the other hand, in case of either medium or low system inertia, the system inertia reduction affects performance, stability, and resiliency of the studied EPS with the conventional control. Therefore, the frequency response of the studied EPS with the conventional control turns into instability once the solar power is connected at $t = 0$ s; thus, the digital OUFR energizes and sends a trip signal to the generator circuit breaker in that time. The proposed virtual inertia control based on VSG can properly maintain the frequency change within the acceptable range of ± 0.012 Hz and ± 0.021 Hz in the circumstances of medium and low system inertia, respectively, as shown in Fig. 5.17b, c. Hence, the proposed coordinated scheme of frequency control including the virtual inertia control based on VSG and the digital OUFR can maintain the dynamic security of the EPS considering high penetration of RESs and stable frequency response compared to the conventional control.

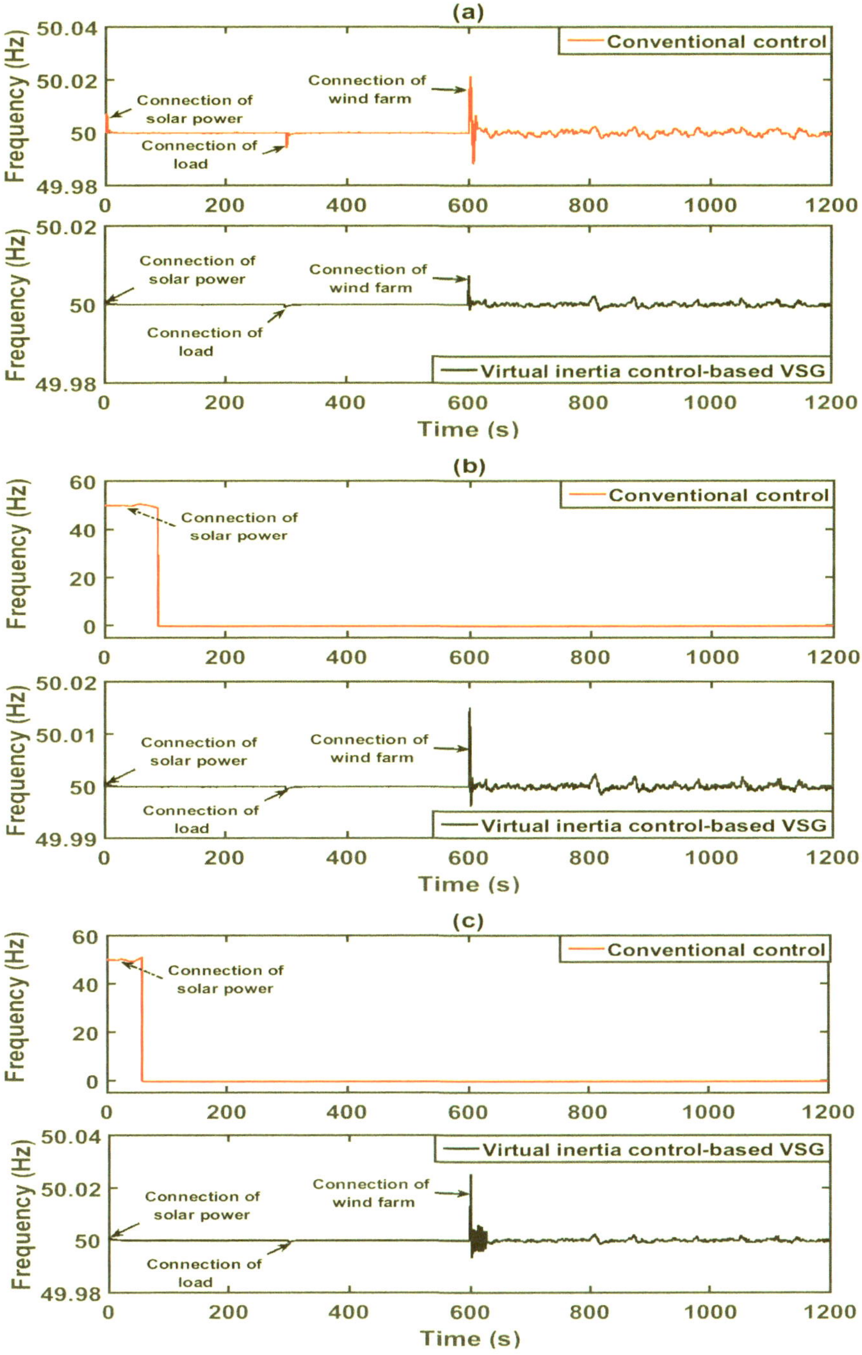

Fig. 5.17 The frequency response of the studied EPS for scenario 3. (**a**) High system inertia, (**b**) medium system inertia, and (**c**) low system inertia

Scenario 4: Multiple Disturbances in Load Demand

In this scenario, the studied EPS with the proposed coordination scheme is investigated in the presence of high-fluctuated wind power and low-fluctuated solar power as shown in Fig. 5.10 considering the assumed various operating states in Table 5.5, and also high-fluctuated random load variation (i.e., industrial loads) and low-fluctuated random load variation (i.e., residential load) as shown in Fig. 5.8 considering the multiple operating conditions of random load variation in Table 5.4. Moreover, the studied EPS with the proposed coordination scheme is tested and evaluated in the circumstances of high, medium, and low system inertia. Figure 5.18 shows the frequency response of the studied EPS under this studied scenario as well as in the circumstances of high, medium, and low system inertia, respectively. From the simulation results, it is clear that conventional control cannot withstand the variation of system frequency. Therefore, the digital OUFR energizes and sends a trip signal to the generator circuit breaker when connecting the wind farm at $t = 600$ s in case of high system inertia (100% of its nominal value), and also when connecting the industrial load at $t = 300$ in cases of medium and low system inertia. The proposed coordination scheme based on the virtual inertia control-based VSG succeeded to restore the frequency to the nominal value at 50 Hz. Therefore, the studied EPS with the proposed coordinated scheme does not need any action from the digital OUFR. Hence, the proposed coordination effectiveness was approved to maintain the dynamic security of the EPS considering high penetration level of the RESs.

5.7 Conclusion

With increasing the penetration level of RESs into power electronics-based RPSs, the overall system inertia might be significantly reduced, increasing the susceptibility of the RPS to the system instability. Thus, maintaining the dynamic security of RPSs has become the main challenge for integration of more RESs. Therefore, this study introduces a new concept of frequency control based on VSG, which includes three virtual models: virtual rotor, virtual primary control, and virtual secondary control to add some inertia to the system control loop virtually and accordingly stabilizing the system frequency during high penetration of RESs. The virtual controller of the VSG is optimally designed using PSO algorithm to produce the optimal control signal to the VSG for emulating the inertia response into RPSs. Furthermore, this study proposes a new coordinated scheme between frequency control loops (i.e., primary and secondary frequency controls) including the virtual inertia control based on VSG and the digital OUFR for enhancement of the frequency stability and preservation of the dynamic security of RPSs because of the high integration level of the RESs. The small-scale power system (e.g., μG) and the large-scale power system (e.g., EPS) are used to validate the effectiveness of the proposed coordination scheme. The simulation results have demonstrated that

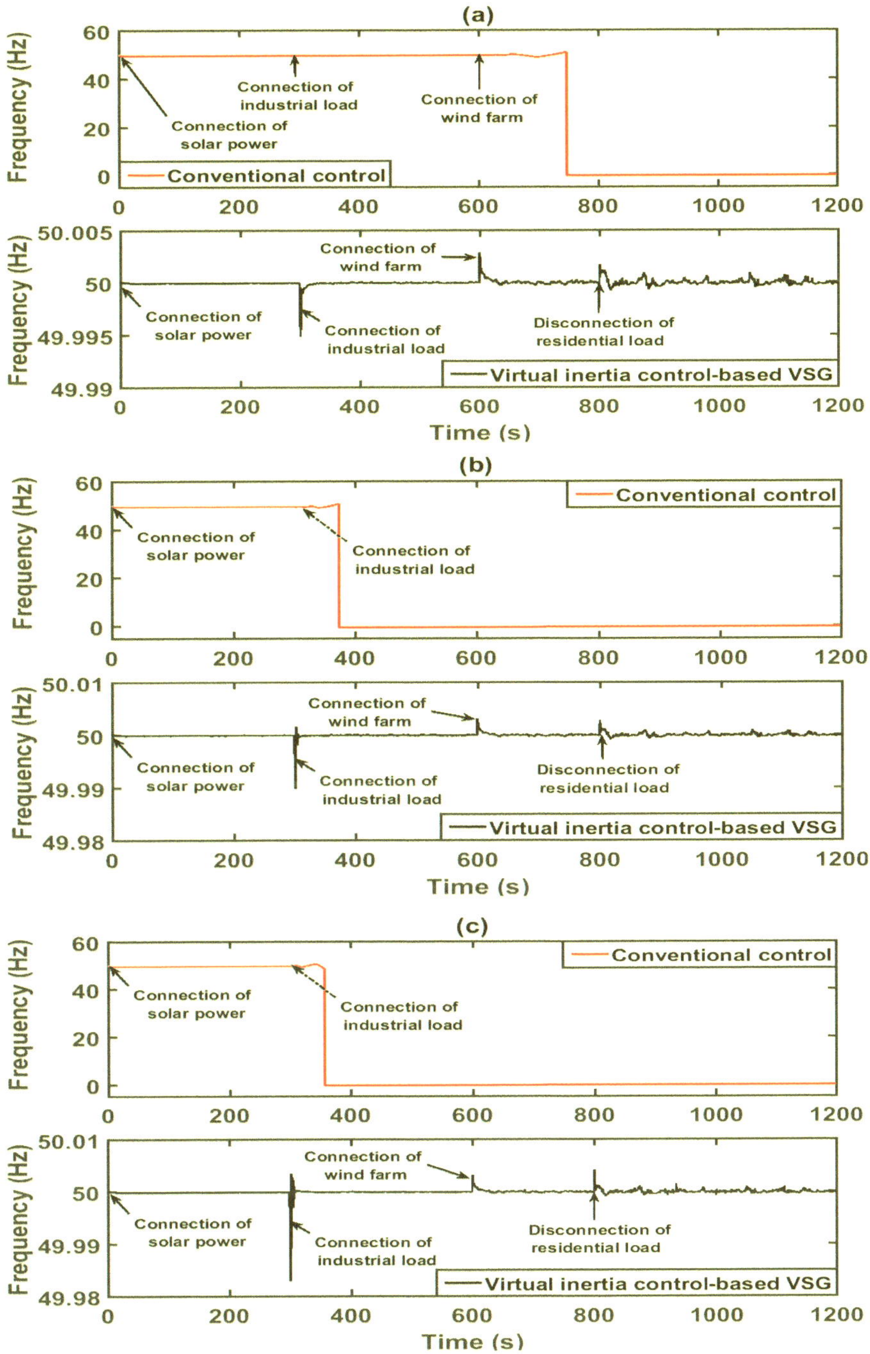

Fig. 5.18 The frequency response of the studied EPS for scenario 4. (**a**) High system inertia, (**b**) medium system inertia, and (**c**) low system inertia

the proposed coordination scheme has achieved effective performance to maintain the dynamic security of RPSs during high penetration level of RESs. However, the proposed virtual inertia control based on VSG has succeeded to restore the system frequency fluctuations to its permissible limits under variation in loading conditions, system inertia (i.e., uncertainty), and nature of the RESs. Also, the results confirmed that the digital frequency protection device has superior accuracy, sensitivity, and extensive control.

Chapter 6
Digital Decentralized Control Scheme in Multi-source Power Systems Based on Mapping Technique

6.1 Introduction

In the past years, several cascaded blackouts happened in electricity networks due to frequency instability. The mismatch between electric power generation and load demand causes a frequency deviation as well as tie-line power deviation in the interconnected power systems. Moreover, the large value of frequency deviation causes many problems such as damaging the equipment, transmission line overloading, and interference with system protection devices [129]. Nowadays, this problem increases after growing the RESs, which have several impacts on the performance of the electrical power systems, such as reduction of the overall system inertia, consequently increasing the frequency and voltage fluctuations [8]. Therefore, the LFC plays a key role in power systems to maintain the system frequency and the power variations at their standard values. To solve the frequency instability problem, numerous control techniques for power system frequency control have been implemented [34, 38, 68–71]. Vrdoljak et al. [38] presented a design of LFC-based sliding mode control for power systems. Yousef et al. [68] applied an adaptive fuzzy logic approach for frequency control of the multi-area interconnected power system. Although the control strategies [38, 68] gave a good dynamic response, they suffer from many drawbacks such as its dependency on the designer's experience and need for long computational time. Ali et al. [69] discussed the robustness of the coefficient diagram method controller including EVs for frequency stability enhancement of a small power system. In addition, Garasi et al. [70] studied the same control strategy in [69] for the LFC of the modern power system. However, the structure of the control technique in [69, 70] is complicated as it requires more steps to get its parameters. Mohamed et al. [71] used decentralized MPC-based LFC in the interconnected power system. Also, Mohamed et al. [34] applied the predictive control strategy for frequency control of interconnected power system concerning wind energy. Although the predictive control strategy in [34, 71] has the advantages of fast response, simple structure, and easy handle system constraints and nonlinearities, it takes more time for

G. Magdy et al., *Renewable Power Systems Dynamic Security*, Power Systems,
https://doi.org/10.1007/978-3-030-33455-0_6

the online calculations at each sampling time. On the other hand, real-world LFC is performed based on PID or PI controllers because they have many merits such as economic cost, simplicity for parameter tuning, robustness, and a successful practical controller, which can provide excellent control performance regardless of the perturbations and variations in the system parameters [72]. However, the PID controller suffers from a complicated process of parameter tuning based on trial-and-error method. In such a case, the robustness of the system is not guaranteed against further perturbations in the system parameters. Therefore, several optimization algorithms were used to find the optimal parameters of the PID controller in the LFC loop such as PSO [6], chaotic optimization algorithm [57], and genetic algorithm [58].

According to the previous studies of the LFC issue, the used model of the most power systems was considered as thermal power plants (e.g., non-reheat and reheat power plants) and/or hydraulic power plants depending on the number of areas [34, 38, 68–71]. However, most of the existing real power systems comprise multisource dynamics generators: thermal, hydraulic, and gas power plants. Therefore, several types of power plants should be added in the LFC problem to achieve a realistic study as reported in this study. And then, this requires using a decentralized control strategy as the dynamic response of each power plant is different than other. Furthermore, the centralized control scheme is difficultly implemented in practice due to the excessive cost of transmitting data over the long distances, as well as the errors, which might be caused accordingly [130]. Hence, the decentralized control scheme is more accurate and realistic than centralized control strategy. On the other side, most of the studied power systems have a linear and simple structure, relying mainly on conventional generators. However, several RESs should be integrated into the interconnected power systems to achieve a more realistic study. Therefore, some studies have addressed the impact of integrating RESs, particularly wind farms, into power systems controlled by various LFC strategies. Hasanien et al. [74] presented a symbiotic organism search algorithm for obtaining the optimal parameters of the frequency controller in the interconnected power system including wind farms. Hasanien [83] used whale optimization algorithm for obtaining the optimal PID controller parameters in an interconnected modern power system including renewable sources. On the other hand, many LFC-based control techniques have neglected the problems associated with communication time delay, which has become one of the most important aspects of uncertainty in LFC analysis because of the complexity of power systems. With the development of power systems, remote signals are available as feedback signals for the design of modern power systems [131]. It is known in control systems that time delay can lead to degradation of system performance and cause even system instability [132]. In light of this fact, the time delay phenomena have been faced in various power systems such as those given in [133, 134]. Moreover, Bevrani et al. [133] proposed a robust decentralized PI controller-based LFC for the time delay power system. Zhu et al. [134] presented a robust coordination scheme between LFC and dynamic demand response for deregulated power systems considering communication delays. Motivated by the aforementioned observations, this study presents a real hybrid power system in Egypt, which includes both conventional generation units (i.e., non-reheat, reheat, and hydraulic power plants) with inherent nonlinearities and wind energy for studying the LFC problem of such a system. Moreover, this study examines the effect of different loading conditions, system parameter variation (i.e., system uncertainty), an uncertainty of wind energy, and time delays on the system performance.

For decades, analog control systems have dominated the control of power system operations. However, increasing the complexity of power systems requires the use of digital devices. Digital control systems that have many advantages such as (a) reduced implementation costs, (b) reduced design time, (c) increased reliability of the control system, and (d) flexibility for system parameter change can be changed once a software program is recompiling [135]. Therefore, digital control systems have become more applicable to replace analog controllers in power systems. Thus, many digital devices are designed and implemented in practical power systems such as digital PID, digital automatic voltage regulator (AVR), and digital PSS [72, 90]. There are two approaches for designing digital control system: (a) direct design approach, which is to discretize the analog system and then to find a suitable digital controller for the discretized system, and (b) indirect design (i.e., redesign approach), which is to design a suitable analog controller for the analog system and then digital redesign of analog control system [136]. This study focuses on the latter approach (i.e., indirect design approach) due to several reasons such as the following: (a) it is more realistic to implement the design of controller in continuous-time mode and (b) it is easy to choose sampling time, which can be selected after defining the continuous-time closed-loop bandwidth. The discrete-time controllers have been implemented for many applications in power systems. To clarify the merits of the proposed digital control model in this study, the features are compared with several related researches [72, 88, 90, 137]. The comparison result is shown in Table 6.1. Based on the above analysis, this study presents a digital model of an optimal PID controller-based PSO algorithm for decentralized LFC in a real hybrid power system (i.e., the EPS) including wind farms. Therefore, the main contribution of this work includes the following aspects.

1. Present a real multisource nonlinear power system (i.e., the EPS), which includes the conventional generation units (i.e., thermal, hydraulic, and gas power plants) with inherent nonlinearities, as well as merging the wind power, which is extracted from Zafarana wind farm, Egypt.
2. Propose a digital model of an optimal PID controller-based PSO algorithm for decentralized LFC of the EPS to replace the traditional analog model for the sake of facing the complexity of modern power systems, reduce the cost of the implementation, and increase the reliability of the control system. Moreover, the proposed digital model of decentralized LFC gives a superior robustness and frequency stabilization effect.

Table 6.1 Comparison of the proposed technique with several related papers

Properties	[90]	[88]	[72]	[137]	This study
Studied power system	Linearized model of single machine infinite bus (SMIB) system			3rd-order simplified model of a linearized power system	Real multisource nonlinear power system including wind farms (i.e., the EPS)
	3rd-order model of a linearized power system	4th-order model of a linearized power system	6th-order model of a linearized power system		
Discretization technique	Digital redesign method (i.e., indirect approach)				
	Bilinear transformation technique	Plant-input-mapping (PIM) technique		Reduced-order PIM technique	Mapping technique
Achieve a desirable performance under all load disturbances by selection:	Very small sampling time (i.e., needs an extreme cost)	Small sampling time (i.e., needs a high cost)			Small and sufficiently large sampling time (i.e., limited reduction of implementation cost)
Effect of changing sampling time intervals					
Selection sampling time based on	Trial and error (i.e., may be a negative effect on costs of hardware requirement)				Sampling theory [138] to obtain optimal sampling time (i.e., saving cost)
Is the controller needs a re-tuning?	Yes				No

3. Investigate the robustness of the proposed digital model against different load patterns, different loading, system parameter variations, an additional wind energy uncertainty, and time delays.
4. Test the admissible ranges of time delays (t_d) and sampling time intervals (T) under the decentralized LFC that remain stable and observe the whole system stability.

6.2 System Configuration and State Equations

6.2.1 Modeling of the EPS

The used model of the EPS has already been described in Chap. 2. Moreover, the system parameter values of the EPS have also been given in Table 2.1. Hence, the simplified model of the studied power system is shown in Fig. 6.1. The nonlinear model of the EPS considering wind energy with the decentralized LFC scheme is represented as shown in Fig. 6.2.

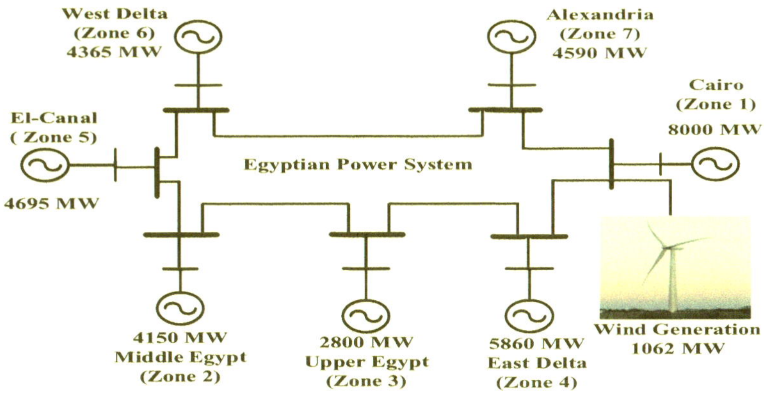

Fig. 6.1 A simplified model of the EPS considering wind farms

6.2.2 Wind Power Generation

According to the Global Wind Energy Council (GWEC), the installed wind power reaches 52.5 GW in 2017. Moreover, the total installed wind power worldwide reaches 539.5 GW at the end of 2017 [139]. Therefore, wind power is the fastest growing and most widely utilized in modern power systems as the cost of installation is lower than other RESs (e.g., PV system). Thus, wind energy represents a significantly larger portion of installed electrical power from RESs. Hence, this study focuses on the effect of integration of wind energy into the EPS on the system frequency stability. The mechanical power from the wind turbine model can be written as the following equations [140]:

$$P_{\text{Wind}} = \frac{1}{2}\rho A_{\text{T}} V_{\text{W}}^{3} C_{\text{P}}(\lambda, \beta) \tag{6.1}$$

where ρ is the air density (kg/m^3), AT is the rotor swept area (m^2), V_{W} is the rated wind speed (m/s), and CP represents the power coefficient of the rotor blades, which can be termed by equations, which are mentioned in Chap. 2. This study uses GAMESA wind turbine, which is installed at Zafarana location in Egypt [141]. Details of this wind turbine are given in Appendix (A).

Also, this study uses real wind speed data at Zafarana location for 1 day, which is extracted from [74]. The rated wind speed is 16 m/s. In this study, the EPS includes a combined model of the wind farm, which has 1250 wind turbine units of 850 kW for each unit besides the conventional generation units (as future planning of the EPS). The wind generator is modeled as a first-order transfer function of a unity gain and 0.3 of a wind time constant (T_{WT}) as shown in Fig. 6.2.

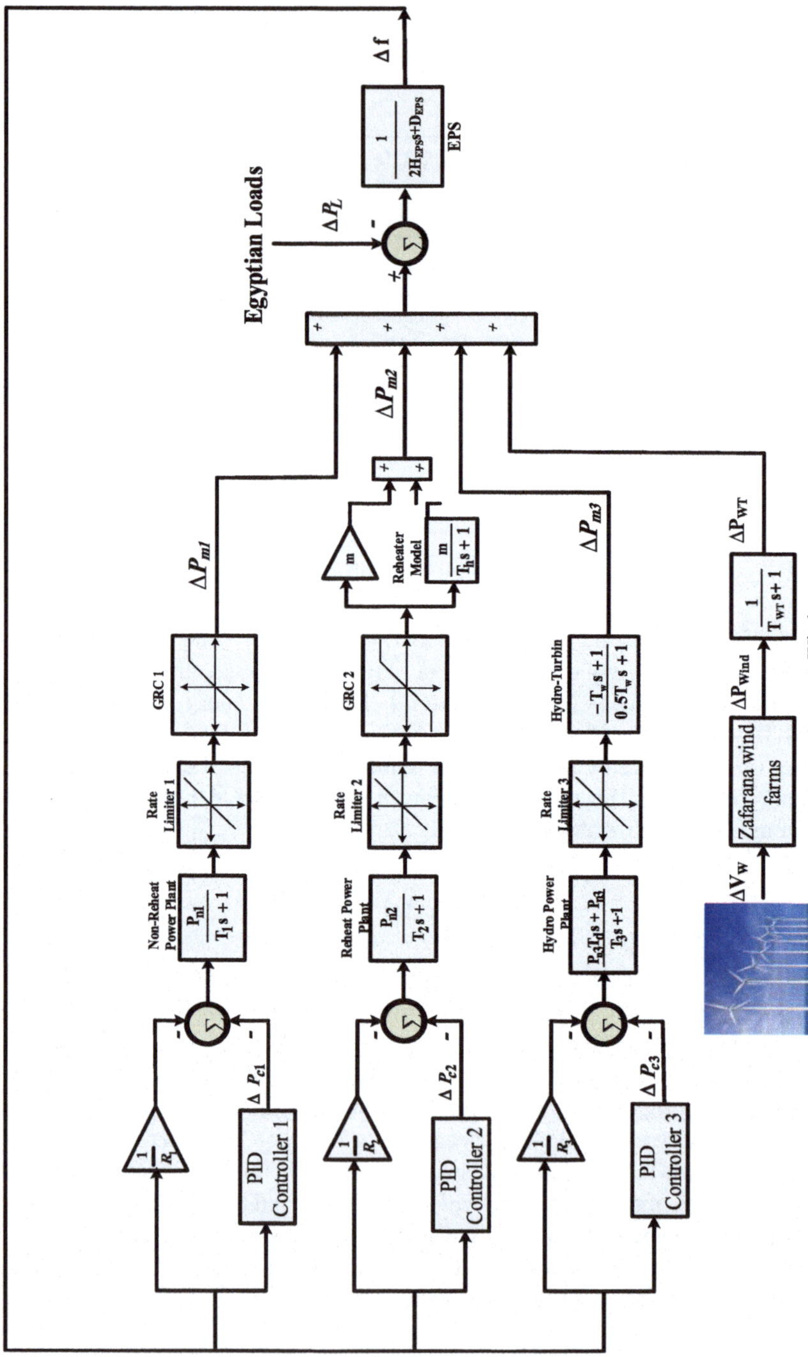

Fig. 6.2 A nonlinear model of the EPS considering wind farms with the decentralized controllers [140]

6.2.3 State-Space Dynamic Modeling

The state-space model is an effective model for the decentralized LFC design in the real hybrid power system synthesis. The frequency deviation of the studied power system considering the effect of the primary frequency control loop (i.e., governor action) and secondary frequency control loop (i.e., LFC) can be obtained as

$$\Delta f = \frac{1}{2H_{\mathrm{EPS}}s + D_{\mathrm{EPS}}}(\Delta P_{m1} + \Delta P_{m2} + \Delta P_{m3} + \Delta P_{\mathrm{WT}} - \Delta P_{\mathrm{L}}) \tag{6.2}$$

Figure 6.2 shows a block diagram of transfer functions describing the different subsystems of the studied nonlinear power system. The blocks are as follows:

1. The non-reheat power plants (e.g., sub-area 1):

$$\Delta P_{m1} = \frac{P_{n1}}{T_1 S + 1} \times \left(\frac{-1}{R_1} \times \Delta f - \Delta P_{\mathrm{c}}\right) \tag{6.3}$$

2. The reheat power plant (e.g., sub-area 2):

$$\Delta P_{g2} = \frac{P_{n2}}{T_2 S + 1} \times \left(\frac{-1}{R_2} \times \Delta f - \Delta P_{\mathrm{c}}\right) \tag{6.4}$$

$$\Delta P_{m2} = \left(m + \frac{m}{T_{\mathrm{h}} S + 1}\right) \times \Delta P_{g2} \tag{6.5}$$

3. The hydropower plants (e.g., sub-area 3):

$$\Delta P_{g3} = \frac{P_{n3} T_{\mathrm{d}} S + P_{n3}}{T_3 S + 1} \times \left(\frac{-1}{R_3} \times \Delta f - \Delta P_c\right) \tag{6.6}$$

$$\Delta P_{m3} = \left(\frac{-T_{\mathrm{w}} S + 1}{0.5 \times T_{\mathrm{w}} S + 1}\right) \times \Delta P_{g3} \tag{6.7}$$

4. The wind turbine model:

$$\Delta P_{\mathrm{WT}} = \frac{1}{T_{\mathrm{WT}} S + 1} \times (\Delta P_{\mathrm{Wind}}) \tag{6.8}$$

In this study, the wind power variation (ΔP_{Wind}) and the load power variation (ΔP_{L}) are considered as the system disturbance signals. (U) is the control input signal and (Y) is the control output signal, which is the system frequency deviation (Δf). Using suitable definitions and state variables from Eqs. (6.2)–(6.8), the linearized state-space model of the EPS from Fig. 6.2 can be easily obtained in the form of the following equations:

$$\dot{X} = \mathrm{AX} + \mathrm{BU} + \mathrm{EW} \tag{6.9}$$

$$Y = \mathrm{CX} + \mathrm{DU} + \mathrm{FW} \tag{6.10}$$

where

$$X^T = \left[\Delta f \;\; \Delta P_{m1} \;\; \Delta P_{m2} \;\; \Delta P_{g2} \;\; \Delta P_{m3} \;\; \Delta P_{g3} \;\; \Delta P_{\mathrm{W}}\right] \tag{6.11}$$

$$U = \left[\Delta P_{\mathrm{C}}\right] \tag{6.12}$$

$$W^T = \left[\Delta P_{\mathrm{Wind}} \Delta P_{\mathrm{L}}\right] \tag{6.13}$$

Hence, the complete state-space equations for the EPS considering wind energy can be obtained as in Eq. (6.14). There are two inputs (ΔP_{L}) and (ΔP_{C}); the output is the frequency deviation (Δf) in the studied system, considering (ΔP_{L}) as load disturbance input in this study. Using the data given above, the transfer function of the EPS model $G(s)$ given by Fig. 6.2 and the state-space equations given by Eq. (6.14) can be calculated using the MATLAB function "SS2TF" in the signal processing toolbox and are given by Eq. (6.15):

$$
\dot{X} = \begin{bmatrix}
-\frac{D_{\mathrm{EPS}}}{2H_{\mathrm{EPS}}} & \frac{1}{2H_{\mathrm{EPS}}} & \frac{1}{2H_{\mathrm{EPS}}} & 0 & \frac{1}{2H_{\mathrm{EPS}}} & 0 & \frac{1}{2H_{\mathrm{EPS}}} \\
-a_1 & -\frac{1}{T_1} & 0 & 0 & 0 & 0 & 0 \\
-ma_2 & 0 & -\frac{1}{T_{\mathrm{h}}} & b_1 & 0 & 0 & 0 \\
-a_2 & 0 & 0 & -\frac{1}{T_2} & 0 & 0 & 0 \\
(-2b_2 D_{\mathrm{EPS}} + 2a_3) & 2b_2 & 2b_2 & 0 & \left(2b_2 - \frac{2}{T_{\mathrm{w}}}\right) & \left(\frac{2}{T_{\mathrm{w}}} + \frac{2}{T_3}\right) & 0 \\
(b_2 D_{\mathrm{EPS}} - a_3) & -b_2 & -b_2 & 0 & -b_2 & -\frac{1}{T_3} & 0 \\
0 & 0 & 0 & 0 & 0 & 0 & -\frac{1}{T_{\mathrm{WT}}}
\end{bmatrix}
\begin{bmatrix} \Delta f \\ \Delta P_{g_1} \\ \Delta p_{m_2} \\ \Delta p_{g_2} \\ \Delta P_{m_3} \\ \Delta P_{g_3} \\ \Delta P_{\mathrm{WT}} \end{bmatrix} +
\begin{bmatrix} 0 \\ -\frac{P_{n_1}}{T_1} \\ -\frac{m * P_{n_2}}{T_2} \\ -\frac{P_{n_2}}{T_2} \\ \frac{2 * P_{n_3}}{T_3} \\ -\frac{P_{n_3}}{T_3} \\ 0 \end{bmatrix} [\Delta P_{\mathrm{C}}] +
\begin{bmatrix} 0 & -\frac{1}{2H_{\mathrm{EPS}}} \\ 0 & 0 \\ 0 & 0 \\ 0 & 0 \\ 0 & -2b_2 \\ 0 & -2b_2 \\ \frac{1}{T_{\mathrm{WT}}} & 0 \end{bmatrix}
\begin{bmatrix} \Delta P_{\mathrm{Wind}} \\ \Delta P_{\mathrm{L}} \end{bmatrix}, \quad Y = [1\,0\,0\,0\,0\,0\,0][\Delta f] \tag{6.14}
$$

The constants of matrices are

$$
a_1 = \frac{P_{n1}}{T_1 R_1}, \quad a_2 = \frac{P_{n2}}{T_2 R_2}, \quad a_3 = \frac{P_{n3}}{T_3 R_3}, \quad a_4 = \frac{T_{\mathrm{d}}}{2H_{\mathrm{EPS}}}, \quad b_1 = \frac{2m}{T_{\mathrm{h}}} - \frac{m}{T_2}, \quad b_2 = a_3 a_4
$$

$$
G(s) = \frac{-0.08757S^5 - 0.6286S^4 - 1.532S^3 - 1.349S^2 - 0.19725S - 0.002027}{S^6 + 7.179S^5 + 17.56S^4 + 15.67S^3 + 2.601S^2 + 0.098035S + 0.001034} \tag{6.15}
$$

6.3 Control Methodology

6.3.1 *Continuous-Time of the Decentralized Control Scheme*

In this study, the proposed decentralized LFC for the EPS considering wind energy is based on the PID controller, which is one of the earliest industrial controllers. The PID controller is composed of three terms (i.e., gains), which are proportional gain K_p, integral gain K_i, and derivative gain K_d. Its transfer function is expressed as follows:

$$G_\mathrm{c}(s) = K_\mathrm{p} + \frac{K_\mathrm{i}}{s} + K_\mathrm{d}s \tag{6.16}$$

According to the suffering of the PID controller to find the optimal parameters, this study uses an intelligent searching method (i.e., PSO). The PSO algorithm as a simple and fast searching intelligent technique has been modeled to tune the parameters of the PID controller. The main objective of the PSO algorithm is to minimize the system frequency deviation by obtaining the optimal PID controller parameters. Moreover, the ISE is used in this study as an objective function of the optimization technique as in Chap. 4. The details of the PSO algorithm have already been given in Chap. 4. Therefore, the optimal gains of the decentralized PID controllers using the PSO are indicated in Table 6.2.

6.3.2 *Discrete-Time of the Decentralized Control Scheme*

In this study, the discretization process of the optimal continuous-time PID controller is implemented using the mapping technique. The discretization process is the process of converting the continuous-time system to discrete-time system. The mapping technique has three classical discretization approaches, which convert a model from S-domain to Z-domain. The three approaches are (a) forward difference method (FDM), (b) backward difference method (BDM), and (c) bilinear transformation method (i.e., Tustin's method) [137, 140]. This study focused on Tustin's discretization method of mapping technique because it has more advantages than other approaches such as ease of implementation and convergence to analog one. Moreover, in this technique, the left-hand side of the *S*-plane is mapped within the unit circle in the *Z*-plane [90]. The discrete-time PID controller approximation is obtained for the transfer function of the continuous-time PID controller simply by replacing S-domain to Z-domain. According to this relation:

Table 6.2 Optimal values of the PSO-based decentralized PID controllers

PID controllers	PID parameter K_p	K_i	K_d
PID controller 1	26.5370	16.3125	−0.5080
PID controller 2	9.68204	0.806941	18.73075
PID controller 3	177.54086	0.5671125	0.1

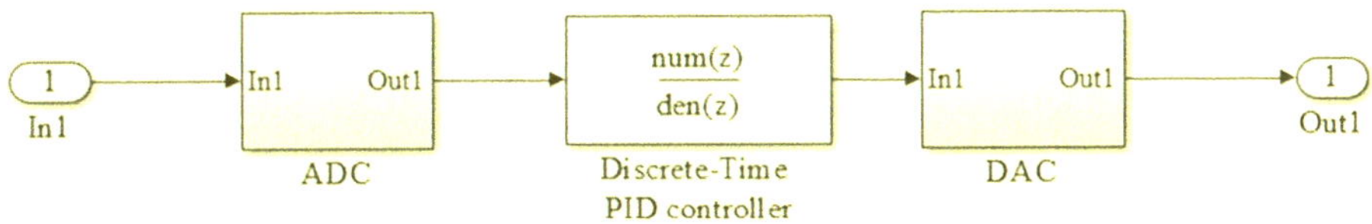

Fig. 6.3 Block diagram of a discrete-time controller with sampling and hold devices

$$s = \frac{1}{T}\ln\ z = \frac{z-1}{\frac{T}{2} \times (z+1)} \tag{6.17}$$

where T is the sampling interval of the discrete-time system. The discrete-time model of the decentralized PID controller for the LFC problem of the EPS is a hybrid system that uses a combination of continuous-time signals and discrete-time signals. The system-controlled signal (i.e., the continuous-time signal (Δf)) is converted to discrete-time signal via analog to digital converter (ADC), which is inserted into the digital computer. The output signal of the controller (i.e., discrete-time signal) is converted to continuous-time signal via digital to analog converter (DAC) to compensate the frequency deviation of the continuous-time system. The digital signal is obtained by sampling the continuous-time signal at regular sampling time and this signal is called a sampled signal. On the other hand, the DAC and hold circuits are used to obtain the continuous-time signal from the digital one. The block diagram of the discrete-time controller with sampling and hold devices is shown in Fig. 6.3.

The selection of sampling time plays a key role in the design of the digital control system. In fact, the selection of sampling intervals influences the cost of the digital hardware, which is one of the important criterions in the design of the power system. Therefore, the sampling rates should be carefully selected, where designing the discrete-time model of the controller with lower sampling rates often gives rise to a decrease in the cost of hardware. However, the selected sampling time must be agreed with the sampling theory [138]. The continuous-time signal is discretized at a rate at least twice the maximum frequency component of the closed-loop system ω_M according to this relation:

$$\omega_s \geq 2\omega_M \tag{6.18}$$

The maximum frequency is defined as $\omega_M = \frac{2\pi}{T}$, where T is the sampling period. The minimum required sampling rate is $\omega_s = 2\omega_M$, which is called Nyquist rate. In this study, the maximum frequency is estimated at the corner frequency point (−3 dB) of the frequency response of the closed-loop system as shown in Fig. 6.4. Therefore, the proposed discrete-time model is based on the following principle: conversion of the continuous-time controller to the discrete-time, and then to digital form at the end of the design phase. Hence, the typical steps to obtain the proposed discrete-time model of decentralized LFC are given in Fig. 6.5.

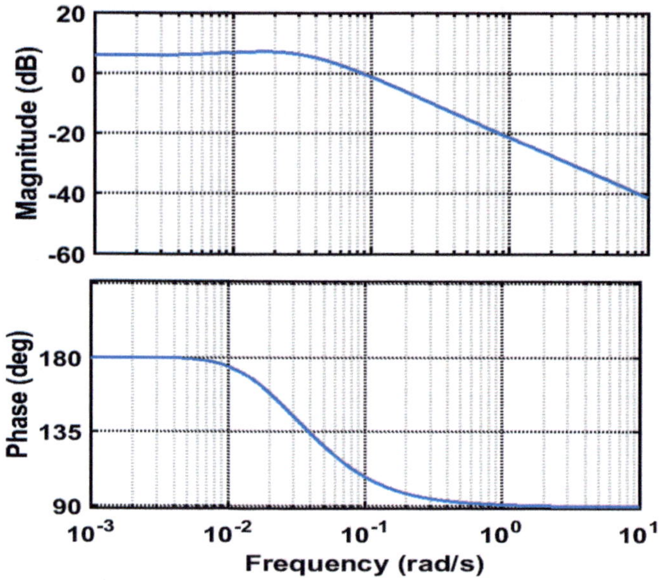

Fig. 6.4 Bode diagram of the closed-loop system

6.4 Simulation Results and Discussion

As a result of technological advances, the control systems have been changing continuously and periodically since the last decade and to this day. Thence, the designers of today's power systems seek to implement digital devices to cope with the increasing complexity of RPSs, thus improving cost and usability [135]. Moreover, digital control is comparatively more accurate, reliable, and flexible; more compact in size; and less sensitive to noise and drift [88]. Consequently, the discrete-time controllers become more appealing to replace analog controllers in different power systems. Hence, we have focused more of our attention on proposing digital decentralized LFC control scheme for the real hybrid power system in Egypt as the first step for upgrading the EPS towards the smart grid.

Therefore, the EPS is taken as the test system to evaluate the performance of the proposed digital control scheme. The model of the EPS considering wind energy is carried out using MATLAB/Simulink® software. Moreover, the GRCs of different generation sources and GDB are taken into account. The code of the PSO algorithm is interfaced as an m-file with the EPS model to execute the optimization process. The performance of the EPS is tested with the proposed digital model of the decentralized LFC optimized and compared to the analog model with variations in load patterns, loading conditions, system parameters, wind farm penetration, and time delay. Therefore, the dynamic response of the EPS is obtained with two control schemes under different scenarios. The proposed digital model of decentralized LFC is designed for pathological sampling intervals at 0.08, 0.1, 0.2, and 0.3 s, respectively. The studied scenarios are as follows:

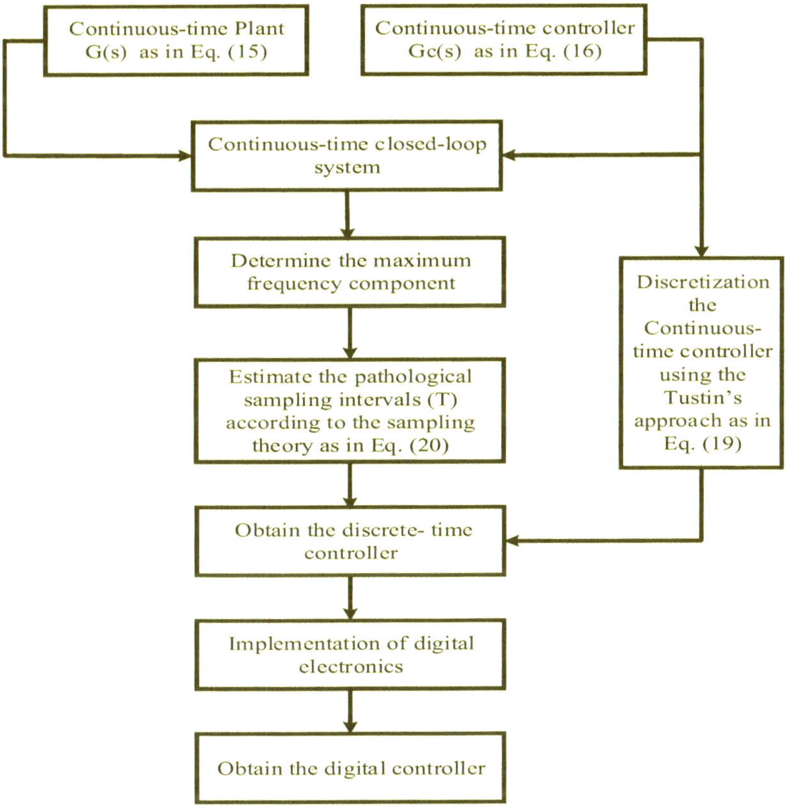

Fig. 6.5 Typical steps for obtaining the proposed digital model [140]

6.4.1 Scenario 1: Robustness Analysis in Case of Real Load Pattern

In this scenario, the EPS with the proposed digital model of the decentralized LFC optimized is tested and evaluated under the influence of a realistic load pattern for a short-term study (i.e., 15 min). The combination of both step load change (e.g., forced outage of the generation unit and the sudden switch to offload) and ramp load change (e.g., industrial load produces a random change in the load) creates a realistic load pattern. Therefore, the realist load is considered as a combination of high random load change (i.e., industrial load), medium random load change (i.e., official load), and low random load change (i.e., residential load). Figure 6.6 shows that the frequency response of the EPS is affected by the random changes of a realistic load. From Fig. 6.6, it has been noticed that the discrete-time model of the decentralized LFC optimized has almost the same effect of the analog model for damping the frequency deviation of the EPS model. On the other hand, in large sampling intervals, the proposed digital model gives a slightly larger oscillation than the

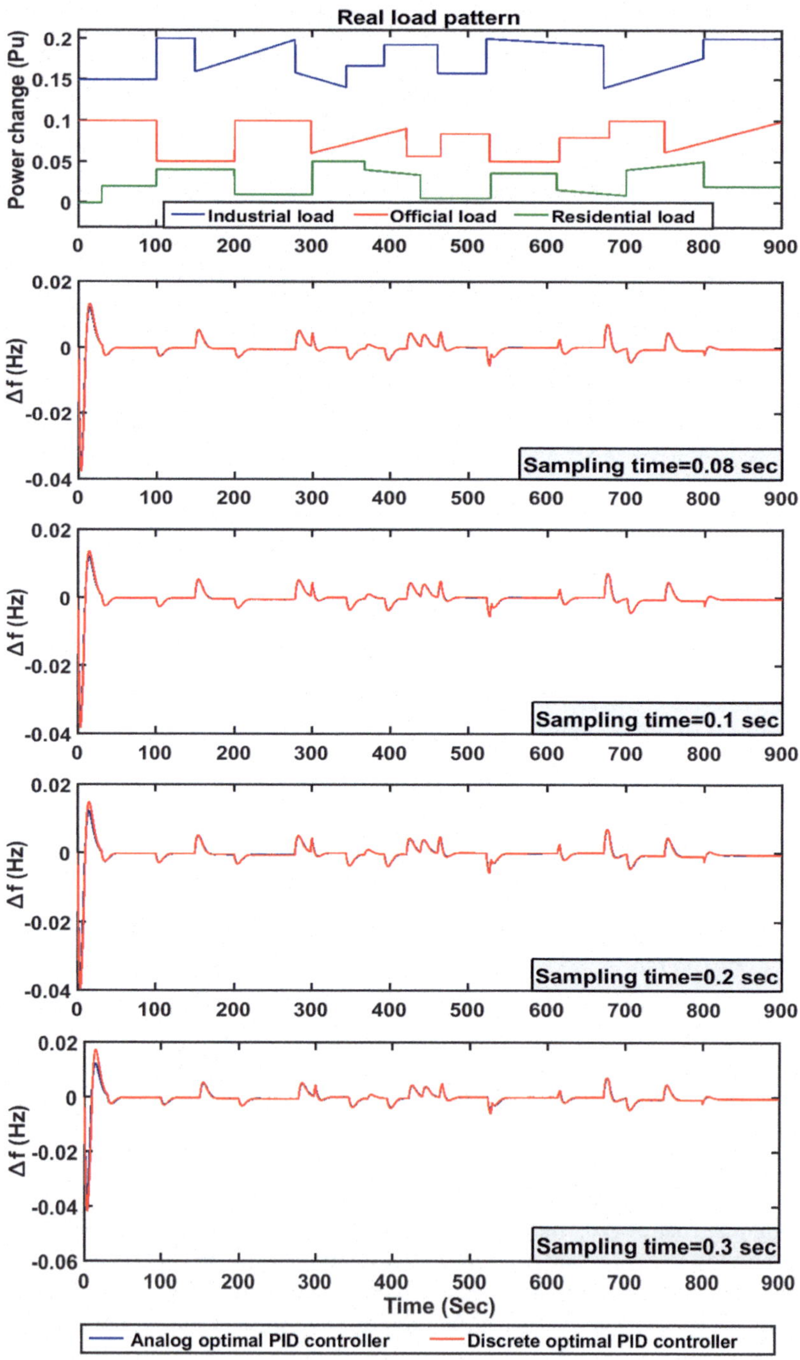

Fig. 6.6 The frequency deviation of the EPS for scenario 1

analog model. However, it gives robust stability in terms of steady-state error and settling time than the analog model. The use of large sampling intervals to design the proposed digital model reduces the cost of hardware requirements. Moreover, there is no need to redesign the PID controllers to deal with the changing of sampling interval process. Hence, this refers to evidence of the robustness of the proposed digital model. The performance specifications, the maximum overshoot (MOS), the maximum undershoot (MUS), and the maximum settling time (T_S) of the studied system, with the two control schemes under the study of scenario 1 during the whole period of simulation (15 min) have been compared in Table 6.3.

6.4.2 Scenario 2: Robustness Analysis in Case of Different Load Conditions

The main objective of this scenario is to investigate the performance of the EPS with the proposed digital model of the decentralized LFC optimized under variation in loading conditions. The decentralized LFC is based on the PID controller, which is optimally designed by the PSO algorithm. Step load perturbations (SLPs) of 3, 5, and 10% pu are applied to the studied power system (i.e., the EPS) at time $t = 100$ s. Figure 6.7 shows the frequency deviation of the EPS with the proposed control schematics under these conditions. It can be noted that the dynamic performance of the EPS with the proposed digital model is very close to that of the analog model, particularly at small sampling intervals. Moreover, the system response using the proposed digital model is faster and has a lower steady-state error than the analog model. Although the performance of the proposed digital model design is slightly different from that of the analog model in the case of large sampling time (i.e., reduction of the hardware cost), it is desirable and has a transient specification within the acceptable frequency range of the system. The transient specifications like MOS, MUS, and T_S increase as SLP increases as indicated in Table 6.4.

6.4.3 Scenario 3: Robustness Analysis in Case of System Parameter Variations

In this scenario, the dynamic performance of the EPS is evaluated with the proposed digital model under variation in system parameters (i.e., system uncertainty) and loading conditions. The variable parameters are T_1, T_2, T_3, T_h, T_d, and T_w, which are changed by $\pm 50\%$ of their nominal values. Furthermore, the different loading conditions (i.e., SLP of 3, 5, and 10% pu) are applied to the EPS as in the previous scenario. Figures 6.8 and 6.9 show the frequency deviation of the studied power system with the proposed two control schemes under these conditions. It is clear from these results that the proposed digital model of the optimal decentralized LFC

Table 6.3 The performance specification of the studied system for scenario 1

	Analog control strategy			Digital control strategy		
Scenario 1	MUS (pu)	MOS (pu)	T_S (s)	MUS (pu)	MOS (pu)	T_S (s)
0.08 s	0.0364	0.0120	22.1	0.0376	0.0128	21.8
0.1 s	0.0364	0.0120	22.1	0.0379	0.0130	22.0
0.2 s	0.0364	0.0120	22.1	0.0390	0.0138	22.1
0.3 s	0.0364	0.0120	22.1	0.0416	0.0142	22.3

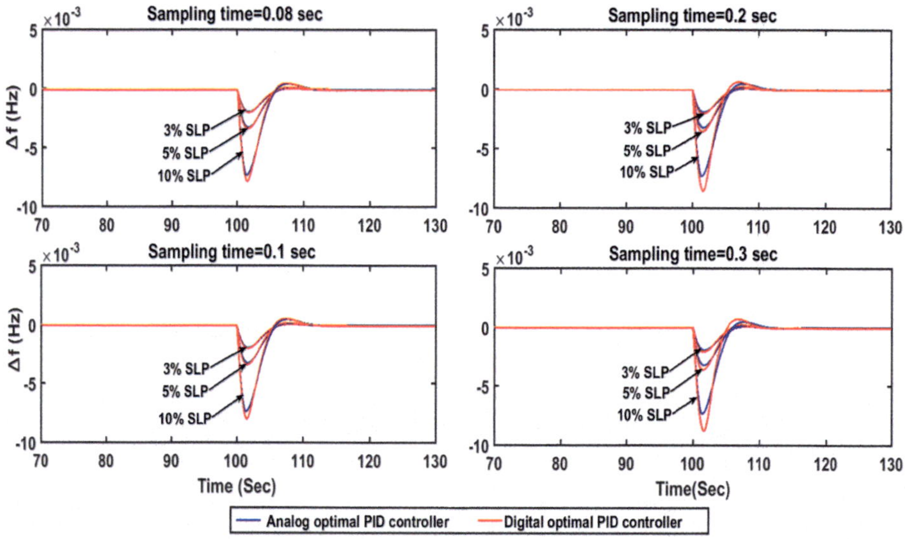

Fig. 6.7 The frequency deviation of the EPS for different load conditions

Table 6.4 The performance specification of the studied system for scenario 2

Scenario 2		Analog control strategy			Digital control strategy		
		MUS (pu)	MOS (pu)	T_S (s)	MUS (pu)	MOS (pu)	T_S (s)
3% SLP	0.08 s	0.00187	0.00010	11.84	0.00190	0.00011	10.32
	0.1 s	0.00187	0.00010	11.84	0.00190	0.00011	10.74
	0.2 s	0.00187	0.00010	11.84	0.00200	0.00013	10.97
	0.3 s	0.00187	0.00010	11.84	0.00200	0.00013	10.76
5% SLP	0.08 s	0.00318	0.00019	11.87	0.00320	0.00019	11.22
	0.1 s	0.00318	0.00019	11.87	0.00330	0.00020	11.43
	0.2 s	0.00318	0.00019	11.87	0.00334	0.00021	11.65
	0.3 s	0.00318	0.00019	11.87	0.00340	0.00022	11.67
10% SLP	0.08 s	0.00728	0.00049	12.22	0.00725	0.00050	11.47
	0.1 s	0.00728	0.00049	12.22	0.00737	0.00050	11.58
	0.2 s	0.00728	0.00049	12.22	0.00802	0.00055	11.31
	0.3 s	0.00728	0.00049	12.22	0.00810	0.00064	11.47

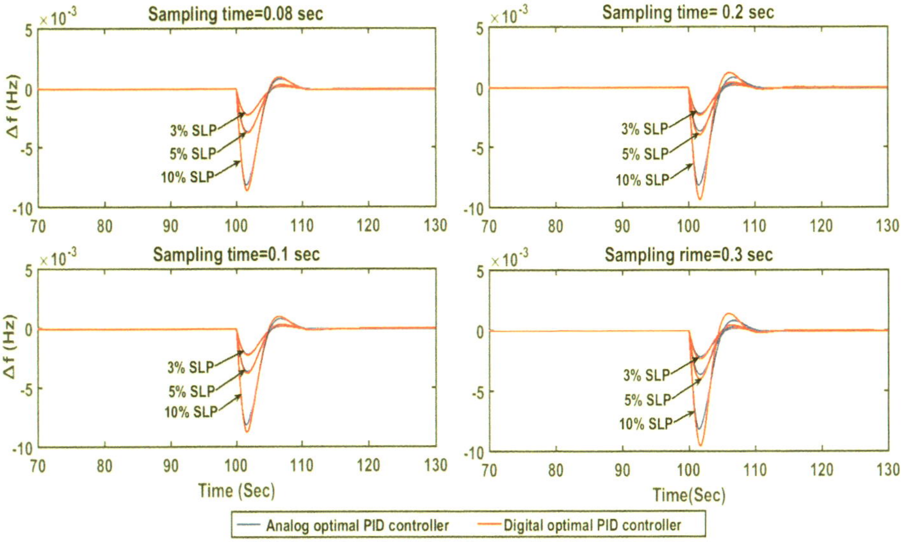

Fig. 6.8 The frequency deviation of the EPS under system parameter variation (+50%)

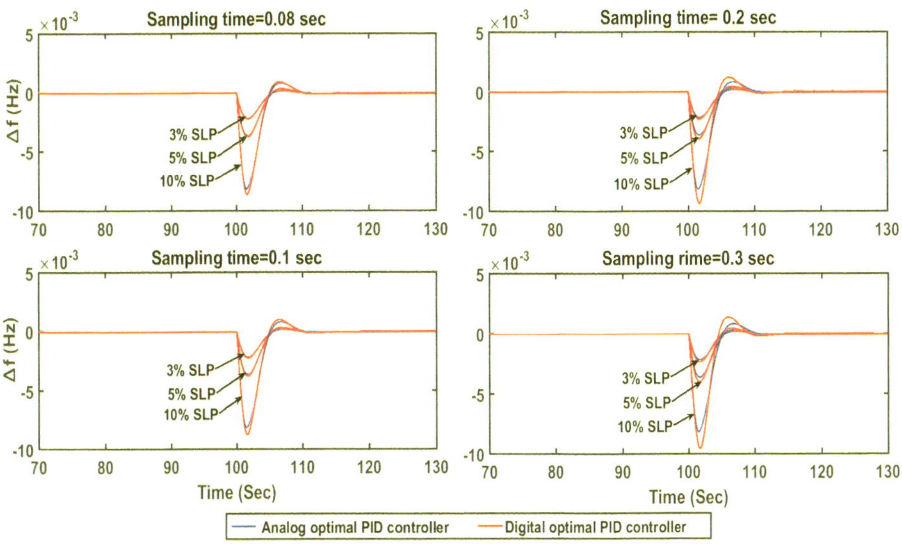

Fig. 6.9 The frequency deviation of the EPS under system parameter variation (−50%)

can effectively regulate the system frequency and ensures robust performance against system uncertainty. Furthermore, the settling time has lower values using the proposed digital model than that by using the analog model. In addition, the numerical results of the transient specification (i.e., MOS, MUS, and T_S) for the two proposed control schemes under variation in loading conditions and system parameters are very close to those of the nominal value of system parameters and within the acceptable ranges of the system frequency as indicated in Table 6.5. Hence, the optimized controller is a robust controller which does not need retuning of its parameters when the EPS is subjected to these conditions. This demonstrates the superiority of the proposed digital model of the decentralized LFC optimized in regulation of the system frequency in case of variation in load conditions and system parameters.

6.4.4 Scenario 4: Robustness Analysis in Case of Uncertainty of Wind Energy

In this scenario, the robustness and the effectiveness of the proposed digital model are tested when the EPS is subjected to wind energy uncertainties. In order to obtain a complete realistic study of the power system in Egypt, a 1062 MW wind farm at the Zafarana location in Egypt is switched on to the EPS at $t = 0$ as shown in Fig. 6.2. This study uses real wind speed data extracted from Zafarana location for 1 day, which has been taken from [74]. Therefore, the actual power output of this wind farm is illustrated in Fig. 6.10, where it fluctuates at its rated value of 0.0301 pu. In addition, it is clear that the real wind power fluctuates randomly due to the nature of wind speed at the Zafarana location. However, as future planning, this study uses three wind farms (1062 MW, 1761 MW, and 3522 MW), which are tested, respectively, to validate the robustness of the proposed digital model to enhance the frequency stability. Moreover, 3% pu of SLP is implemented to the EPS at time $t = 0$. Both the proposed control schemes for this simulation are considered as a 24-h long-term study, and system frequency results are shown in Fig. 6.11. From the simulation results, it is obviously clear that the proposed digital model of the optimal decentralized LFC can suppress the frequency fluctuations even with a change of wind power penetration level for all the pathological sampling intervals. Therefore, the proposed digital model enables the integration of a high level of wind energy in the EPS. The numerical results of the transient specification (i.e., MOS, MUS, and T_S) for the two proposed control schemes are indicated in Table 6.6. These numerical results show that, although the proposed two control schemes have a very close transient specification to some extent, the response of the proposed digital model is faster and has a lower steady-state error for all the pathological sampling intervals than the analog model.

Table 6.5 The performance specification of the studied system for scenario 3

Scenario 3			Analog control strategy			Digital control strategy		
			MUS (pu)	MOS (pu)	T_S (s)	MUS (pu)	MOS (pu)	T_S (s)
Case A (+50%)	3%	0.08 s	0.00214	0.00019	10.76	0.00215	0.00021	10.54
		0.1 s	0.00214	0.00019	10.76	0.00221	0.00021	10.32
		0.2 s	0.00214	0.00019	10.76	0.00228	0.00023	9.995
		0.3 s	0.00214	0.00019	10.76	0.00231	0.00026	9.477
	5%	0.08 s	0.00360	0.00033	10.78	0.00366	0.00033	10.34
		0.1 s	0.00360	0.00033	10.78	0.00370	0.00033	10.13
		0.2 s	0.00360	0.00033	10.78	0.00379	0.00034	9.578
		0.3 s	0.00360	0.00033	10.78	0.00388	0.00039	9.199
	10%	0.08 s	0.00807	0.00082	10.93	0.00809	0.00088	10.06
		0.1 s	0.00807	0.00082	10.93	0.00813	0.00090	9.848
		0.2 s	0.00807	0.00082	10.93	0.00854	0.00093	9.457
		0.3 s	0.00807	0.00082	10.93	0.00887	0.00098	9.239
Case B (−50%)	3%	0.08 s	0.00161	0.00005	12.93	0.00162	0.00005	12.06
		0.1 s	0.00161	0.00005	12.93	0.00166	0.00005	11.63
		0.2 s	0.00161	0.00005	12.93	0.00169	0.00006	9.891
		0.3 s	0.00161	0.00005	12.93	0.00178	0.00007	9.022
	5%	0.08 s	0.00275	0.00010	12.30	0.00284	0.00010	11.65
		0.1 s	0.00275	0.00010	12.30	0.00284	0.00011	11.22
		0.2 s	0.00275	0.00010	12.30	0.00300	0.00013	11.00
		0.3 s	0.00275	0.00010	12.30	0.00305	0.00014	10.78
	10%	0.08 s	0.00622	0.00031	12.22	0.00630	0.00031	11.36
		0.1 s	0.00622	0.00031	12.22	0.00656	0.00033	11.58
		0.2 s	0.00622	0.00031	12.22	0.00694	0.00035	11.35
		0.3 s	0.00622	0.00031	12.22	0.00713	0.00029	10.93

6.4.5 Scenario 5: Robustness Analysis in Case of Time Delays

In control systems, particularly digital control, the communication delays are to become a considerable problem as one of the important uncertainties in the LFC analysis due to the complexity of the power system. It is well known that time delays can lead to system degradation and even destabilize the system [133]. Therefore, this study tests the validity and effectiveness of the proposed digital model of the optimal decentralized LFC in the EPS versus time-varying delays. The communication delays are considered on the control input, which is the deviation in measured frequency.

The delays on the measured frequency deviation from remote terminal units to the control center can be transferred to the control signal side as shown in Fig. 6.12. The communication delay is expressed as an exponential function (e^{-st_d}) where (t_d) is the communication delay time. In this scenario, the EPS is tested in the presence of

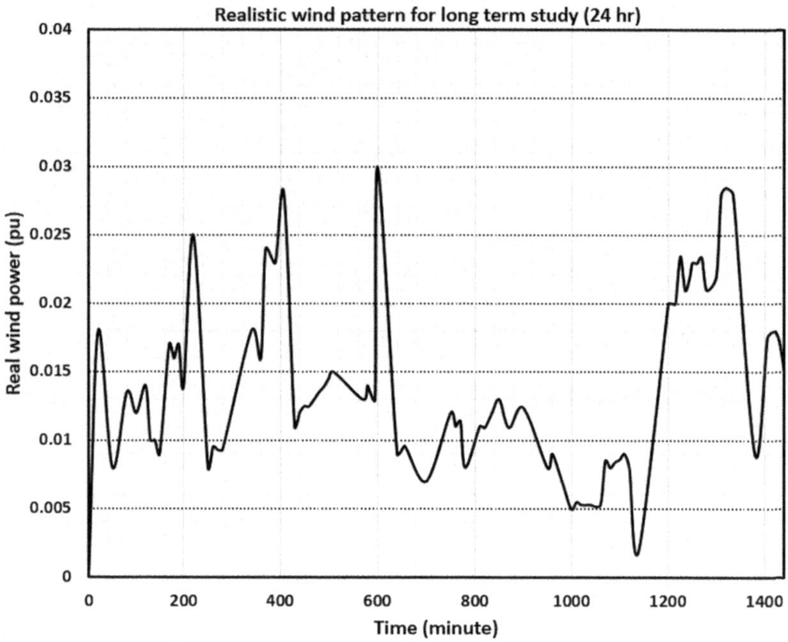

Fig. 6.10 Dynamic response of real power of the wind farm

assumed communication delays for every subsystem that is 0.2 s and a 3% pu of SLP at time $t = 0$. The frequency deviation of the EPS with the proposed control schemes is shown in Fig. 6.13. In this scenario, the EPS is tested in the presence of assumed communication delays for every subsystem that is 0.2 s and a 3% pu of SLP at time $t = 0$. The frequency deviation of the EPS with the proposed control schemes is shown in Fig. 6.13. From the simulation results, both control schemes act to return the frequency deviation to zero for all pathological sampling times for the digital control design. Although the system performance with the proposed digital mode is slightly different from that with the analog model in the case of large sampling intervals, it is faster than the analog model where the frequency deviation quickly reaches to steady-state value. From the robustness test mentioned above, it is found that there is no need to redesign the PID controller to handle communication delay. This refers to evidence of the robustness of the proposed control design. The transient specification (i.e., MOS, MUS, and T_S) for the two proposed control schemes is indicated in Table 6.7.

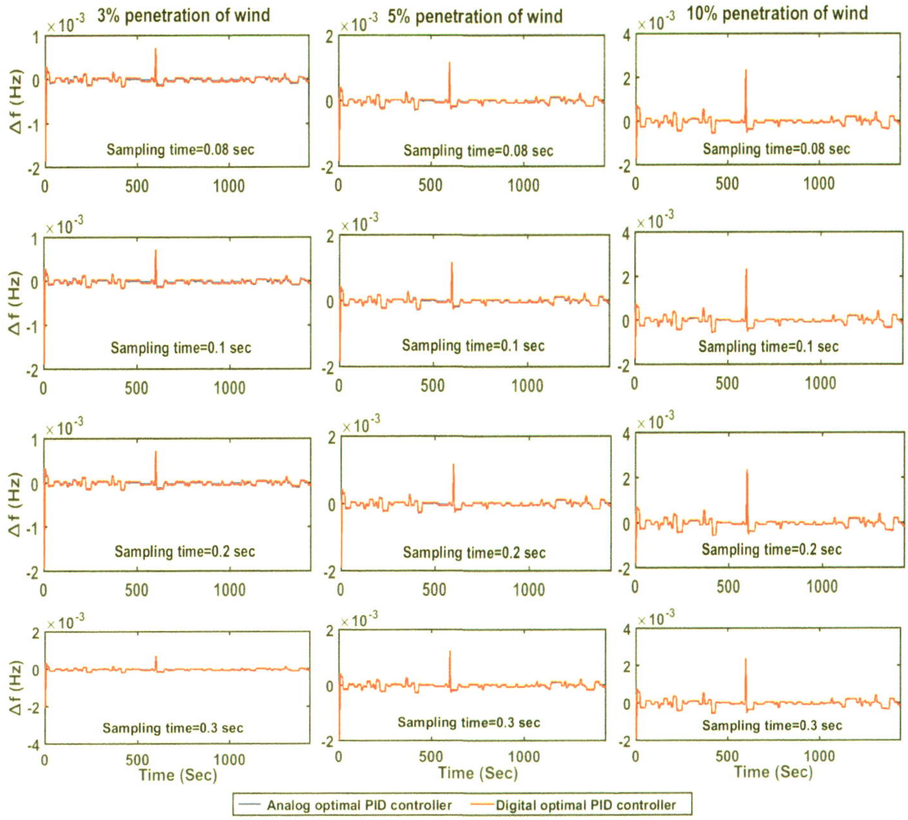

Fig. 6.11 The frequency deviation of the EPS under different wind power penetration levels

Table 6.6 The performance specification of the studied system for scenario 4

Scenario 4		Analog control strategy			Digital control strategy		
		MUS (pu)	MOS (pu)	T_S (s)	MUS (pu)	MOS (pu)	T_S (s)
3% Wind	0.08 s	0.0018	0.0007	22.90	0.00185	0.000718	22.67
	0.1 s	0.0018	0.0007	22.90	0.00188	0.000716	22.35
	0.2 s	0.0018	0.0007	22.90	0.00190	0.000725	22.37
	0.3 s	0.0018	0.0007	22.90	0.00200	0.000728	21.98
5% Wind	0.08 s	0.0017	0.0011	22.94	0.00180	0.001177	22.35
	0.1 s	0.0017	0.0011	22.94	0.00183	0.001179	22.36
	0.2 s	0.0017	0.0011	22.94	0.00189	0.001191	22.53
	0.3 s	0.0017	0.0011	22.94	0.00190	0.001205	22.12
10% Wind	0.08 s	0.0016	0.0023	23.05	0.00168	0.002339	22.61
	0.1 s	0.0016	0.0023	23.05	0.00170	0.002364	22.54
	0.2 s	0.0016	0.0023	23.05	0.00177	0.002389	22.53
	0.3 s	0.0016	0.0023	23.05	0.00180	0.002400	22.22

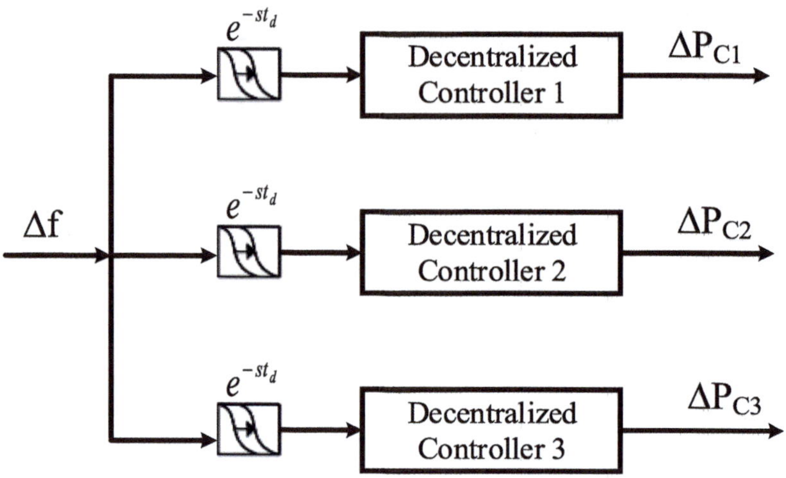

Fig. 6.12 Representation of communication delays in the EPS [140]

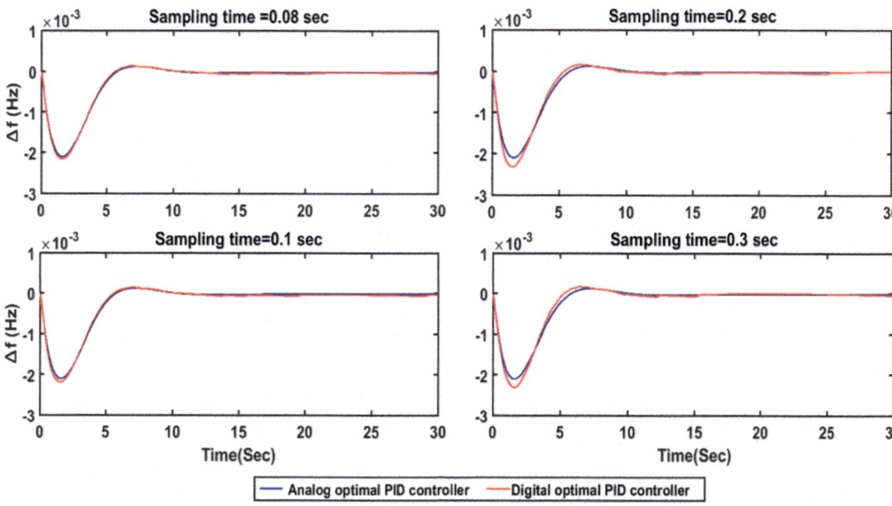

Fig. 6.13 The frequency deviation of the EPS with time delay of 0.2 s for every subsystem

Table 6.7 The performance specification of the studied system for scenario 5

	Analog control strategy			Digital control strategy		
Scenario 5	MUS (pu)	MOS (pu)	T_S (s)	MUS (pu)	MOS (pu)	T_S (s)
0.08 s	0.00208	0.00013	10.95	0.00210	0.00013	9.96
0.1 s	0.00208	0.00013	10.95	0.00217	0.00014	9.62
0.2 s	0.00208	0.00013	10.95	0.00229	0.00014	9.47
0.3 s	0.00208	0.00013	10.95	0.00230	0.00015	9.55

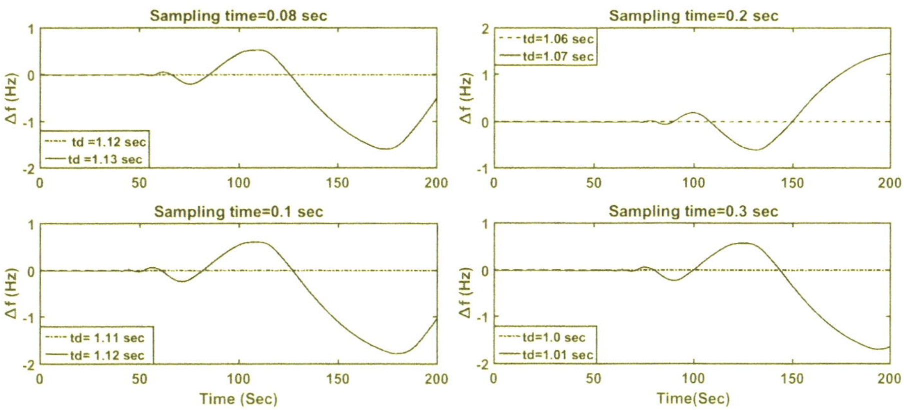

Fig. 6.14 The frequency deviation of the EPS with time-varying delay for different sampling periods

Table 6.8 Stable regions of time-varying delay for the EPS with different sampling periods

Stability index for time-varying delay	
Sampling times	Stable region
0.08 s	[0:1.12]
0.1 s	[0:1.11]
0.2 s	[0:1.06]
0.3 s	[0:1.00]

6.4.6 *Evaluation of System Stability*

6.4.6.1 System Stability Index for Time-Varying Delays

This section tests the admissible range of communication delay, which maintains the state of system stability by manually increasing step by step for the time delay and observing the whole system stability. In this section, the EPS stability index is tested with the proposed digital control scheme under the influence of time delay for different sampling periods. Moreover, 3% pu of SLP is applied to the EPS at time $t = 0$. Figure 6.14 shows the frequency deviation of the EPS with the proposed digital control scheme with a time-varying delay for different sampling periods. The stability regions of time delays for the EPS with the proposed digital control model for different sampling periods are listed in Table 6.8.

6.4.6.2 System Stability Index for Pathological Sampling Intervals

This section tests the admissible range of sampling intervals for the design of the digital control system. It is well known that the selection of sampling intervals is an important issue for economic reasons, as sampling rates should be as low as possible. The low rate means having more time to implement the control algorithm, which can

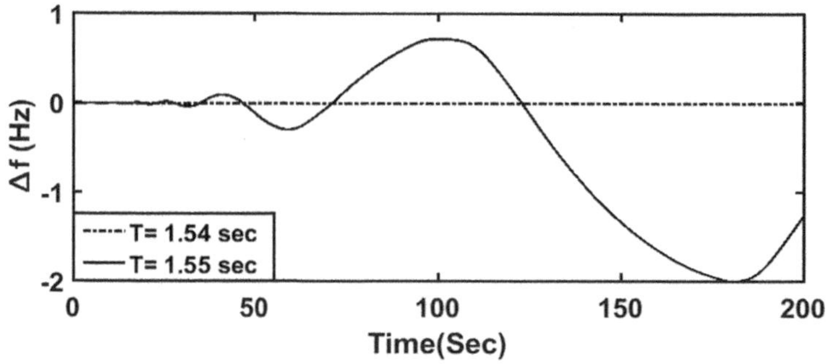

Fig. 6.15 The frequency deviation of the EPS with time delay of 0.2 s for every subsystem for different sampling times

be performed on slower computers. Hence, the cost of the hardware requirement is reduced. However, according to the sampling theory [138], the sampling frequency should be at least twice the bandwidth of the closed-loop system as in Eq. (6.18). On the other hand, pathological sampling rates affect the controllability and observability properties of a continuous-time plant. With a pathological sampling frequency, there may be uncontrollable states, thence preventing the design of state feedback controllers and state estimators [137]. Also, it may cause hidden oscillations in the system response. Therefore, this section tests the admissible range of pathological sampling in which the digital model of the optimal decentralized LFC remains stable, by manually increasing step by step the sampling time and observing the whole system stability. Moreover, the EPS with the proposed digital control scheme is tested in the presence of assumed time delay for every subsystem that is 0.2 s and a 3% pu of SLP at time $t = 0$. Figure 6.15 shows the frequency deviation of the EPS with the proposed digital model under different sampling times. From this test, it is found that the stability region of pathological sampling time is up to 1.54 s.

6.5 Conclusion

This study proposes to replace the continuous-time decentralized LFC model of the real multisource power system (i.e., the EPS) considering communication delays through its digital one. The digital model has almost the same performances as the analog model. Moreover, it reduces the implementation cost and increases the reliability of the control system. The EPS includes both conventional generation units (i.e., non-reheat, reheat, and hydraulic power plants) with inherent nonlinearities (i.e., GDB and GRC) and wind power, which is extracted from the Zafarana wind farm, located in Egypt. The decentralized LFC has been designed using the PID controller, which is optimally designed by the PSO algorithm. Therefore, the optimal digital controller-based Tustin's technique is designed for every subsystem of the EPS separately to guarantee the stability of the overall closed-loop system.

Furthermore, the proposed digital model has been designed for different sampling intervals; and the effect of changing sampling intervals on the performance of the digital model is examined through simulations. The performance of the proposed digital model is tested and compared with the analog model under variation in load patterns, load conditions, system parameters, wind farm penetration, and communication delays. The simulation results reveal that the proposed digital model gives a reliable performance and an effective control satisfactory for different pathological sampling intervals, while it presents a very powerful control strategy for frequency stability enhancement and becomes a very promising tool for facing the problems of model power systems considering high renewable energy penetration.

Chapter 7
Conclusions and Future Work

7.1 Conclusions

In view of the analysis and investigations presented, the main conclusions can be summarized as follows:

1. This book has proposed new frequency control techniques based on SMES system, virtual inertia control, and VSG for frequency stability enhancement of RPS considering high penetration level of RESs.
2. A new coordination of the secondary frequency control (i.e., LFC) and the SMES technology using a new optimal PID controller-based MSA has been proposed to enhance the frequency stability of the EPS concerning HWPP under the impact of different load profiles, random load variation, wind power fluctuations, and system uncertainties.
3. A suitable inertia control technique could be applied together with ESSs to emulate additional inertia power to the community or power system, improving system inertia and eliminating the stability issues. Therefore, this book has proposed a novel concept of frequency control incorporating a virtual inertia controller to emulate virtual inertia into the RPS control loop, thus stabilizing RPS frequency during high penetration of RESs.
4. This book has proposed a new frequency control strategy based on VSG, which emulates the characteristics of a real synchronous generator that are as follows: the inertia and damping properties through the concept of virtual rotor as well as frequency control loops (i.e., primary and secondary frequency control) through the virtual primary and secondary control, to compensate the reduction in system inertia that results from adding more RESs (e.g., non-inertia sources), thus stabilizing the system frequency during high penetration of RESs.
5. The proposed frequency control strategies have coordinated with digital OUFR in RPSs for improvement of the frequency stability and preservation of the dynamic security of RPSs because of the high integration level of the RESs.

G. Magdy et al., *Renewable Power Systems Dynamic Security*, Power Systems,
https://doi.org/10.1007/978-3-030-33455-0_7

6. This book has proposed a digital model of the LFC scheme for the power system to replace the traditional analog model for the sake of facing the complexity of modern power systems, reduce the cost of the implementation, and increase the reliability of the control system.
7. The simulation results of the studied RPSs (e.g., μG and the EPS) have been carried out using MATLAB/Simulink® software to validate the effectiveness of the proposed coordination schemes. Results approved that the proposed coordination schemes can effectively regulate the RPS frequency and guarantee robust performance to preserve the dynamic security of RPS with high penetration of RESs for different contingencies.

7.2 Future Work

Based on this book research, some research points require further future research work, as follows:

1. Design and development of the VSG model considering the effects of high penetration of RESs to support 100% RES penetration. Hence, in an effort to achieve 100% RES penetration, VSG will be considered as one of the necessary parts for today's and future power systems.
2. Upgrading the conventional power systems (e.g., the EPS) towards smart grids.
3. Towards 100% renewable community using VSG.

Appendix A

Wind turbine model: manufacturer: GAMESA (Spain), model: G52/850, rated power: 850 kW, rotor diameter: 52 m, swept area: 2.124 m^2, cut-in wind speed: 4 m/s, rated wind speed: 16 m/s, cutoff wind speed: 25 m/s, maximum generator output speed: 1900 rpm, and output voltage: 690 V.

G. Magdy et al., *Renewable Power Systems Dynamic Security*, Power Systems,
https://doi.org/10.1007/978-3-030-33455-0

Publications from This Book

International Journals

[J1]. **Gaber Magdy**, G. Shabib, Adel A. Elbaset, Yasunori Mitani, "Renewable power systems dynamic security using a new coordination of frequency control strategy based on virtual synchronous generator and digital frequency protection," ***International Journal of Electrical Power and Energy Systems***, **July 2019**, Volume 109, pp. 351–368.

[J2]. **Gaber Magdy**, G. Shabib, Adel A. Elbaset, Thongchart Kerdphol, Yaser Qudaih, Yasunori Mitani, "Decentralized optimal LFC for a real hybrid power system considering renewable energy sources," ***Journal of Engineering Science and Technology***, **April 2019,** Volume 14, Issue 2, pp. 682–697.

[J3]. **Gaber Magdy**, G. Shabib, Adel A. Elbaset, Thongchart Kerdphol, Yaser Qudaih, Hassan Bevrani, Yasunori Mitani, "Tustin's technique based digital decentralized load frequency control in a realistic multi power system considering wind farms and communications delays," ***Ain Shams Engineering Journal***, **February 2019**, DOI: 10.1016/j.asej.2019.01.004

[J4]. **Gaber Magdy**, G. Shabib, Adel A. Elbaset, Yasunori Mitani, "A Novel Coordination Scheme of Virtual Inertia Control and Digital Protection for Microgrid Dynamic Security Considering High Renewable Energy Penetration," ***IET Renewable Power Generation*, 25 February 2019**, Volume 13, Issue 3, pp. 462–474.

[J5]. Hossam Ali, **Gaber Magdy**, Binbin Li, G. Shabib, Adel A. Elbaset, Dianguo Xu, Yasunori Mitani, "A New Frequency Control Strategy in an Islanded Microgrid Using Virtual Inertia Control-Based Coefficient Diagram Method," **IEEE Access**, **January 2019**, Volume 7, pp. 16979–16990.

[J6]. **Gaber Magdy**, G. Shabib, Adel A. Elbaset, Yasunori Mitani, "Optimized coordinated control of LFC and SMES to enhance frequency stability of a real multi-source power system considering high renewable energy penetration,"

G. Magdy et al., *Renewable Power Systems Dynamic Security*, Power Systems,
https://doi.org/10.1007/978-3-030-33455-0

***Protection and Control of Modern Power Systems*, December 2018**, Volume 3, Issue 1, pp. 1–15.

[J7]. **G. Magdy**, Emad Mohamed, G. Shabib, Adel A. Elbaset, Yasunori Mitani, "Microgrid dynamic security considering high penetration of renewable energy," ***Protection and Control of Modern Power Systems***, **December 2018**, Volume 3, Issue 1, pp. 1–11.

[J8]. **Gaber Magdy**, Emad Mohamed, G. Shabib, Adel A. Elbaset, Yasunori Mitani, "SMES Based a New PID Controller for Frequency Stability of a Real Hybrid Power System Considering High Wind Power Penetration," ***IET Renewable Power Generation***, **August 2018**, Volume 12, Issue 11, pp. 1304–1313.

[J9]. Emad A. Mohamed, **Gaber Magdy**, G. Shabib, Adel A. Elbaset, Yasunori Mitani, "Digital coordination strategy of protection and frequency stability for an islanded microgrid," ***IET Generation, Transmission & Distribution***, **August 2018**, Volume 12, Issue 15, pp. 3637–3646.

[J10]. **Gaber Magdy**, G. Shabib, Adel A. Elbaset, Yasunori Mitani, "Frequency Stabilization of Renewable Power Systems Based on MPC With Application to The Egyptian Grid," ***IFAC-PapersOnLine*,January 2018,** Volume 51, Issue 28, pp. 280–285.

[J11]. **G. Magdy**, G. Shabib, Adel A. Elbaset, Thongchart Kerdphol, Yaser Qudaih, Hassan Bevrani, Yasunori Mitani, "A Novel Design of Decentralized LFC to Enhance Frequency Stability of Egypt Power System Including Wind Farms," ***International Journal on Energy Conversion*, January 2018**, Volume 6, Issue 1, pp. 17–29.

International Conferences

[C1]. **G. Magdy**, G. Shabib, Adel A. Elbaset, Yaser Qudaih, Yasunori Mitani, "A Robust Control Strategy for Mitigating Renewable Energy Fluctuations in a Real Hybrid Power System Combined With SMES," ***AIP Conference Proceedings***, **2018**, Volume 1968, Issue 1, pp. 1–12.

[C2]. **G. Magdy**, Emad A. Mohamed, G. Shabib, Adel A. Elbaset, and Yasunori Mitani "Enhancement LFC of a Realistic Multi-Source Power System Concerning Wind Farms Using SMES and New Optimized PID Controller," ***IEEE Conference***, 5th International Conference on Electric Power and Energy Conversion Systems (EPECS), Kitakyushu, Japan, **2018**, pp. 1–7.

[C3]. **G. Magdy**, Emad A. Mohamed, G. Shabib, Adel A. Elbaset, and Yasunori Mitani "A Novel Optimal LFC in a Real Hybrid Power System Incorporating SMES Technology," ***IEEE Conference***, International Conference on Electrical, Electronics, Computers, Communication, Mechanical and Computing (EECCMC), Priyadarshini Engineering College, India, 28-29 January **2018**, pp. 1–7.

[C4]. **G. Magdy**, Abualkasim Bakeer, G. Shabib, Adel A. Elbaset, and Yasunori Mitani "Discrete-time optimal controller for load frequency control of multi-source power system in Egypt," ***IEEE Conference***, International Conference on Innovative Trends in Computer Engineering (ITCE), Aswan University, Egypt, February 19–21, **2018**, pp. 264–270.

[C5]. **G. Magdy**, Abualkasim Bakeer, G. Shabib, Adel A. Elbaset, and Yasunori Mitani "Decentralized Model Predictive Control Strategy of a Realistic Multi Power System Automatic Generation Control," ***IEEE Conference***, 19th International Middle East Power Systems Conference (MEPCON'19), Menoufia University, Egypt, December 19–21, **2017**, pp. 190–196.

[C6]. **G. Magdy**, G. Shabib, Adel A. Elbaset, Yaser Qudaih and Yasunori Mitani, "Upgrading power system in Egypt towards smart grid," ***IEEE Conference***, International Conference on Advanced Control Circuits Systems (ACCS) Systems & 2017 International Conference on New Paradigms in Electronics & Information Technology (PEIT), Alexandria, Egypt, **2017**, pp. 251–263.

References

1. H. Bevrani, M. Watanabe, Y. Mitani, *Power system monitoring and control* (John Wiley & Sons, Hoboken, NJ, 2014)
2. E. Rakhshani, D. Remon, A.M. Cantarellas, P. Rodriguez, Analysis of derivative control based virtual inertia in multi-area high-voltage direct current interconnected power systems. IET Gener. Transm. Distrib. **10**(6), 1458–1469 (2016)
3. S. Teimourzadeh, F. Aminifar, M. Davarpanah, Microgrid dynamic security: challenges, solutions and key considerations. Electr. J. **30**(4), 43–51 (2017)
4. N.E.Y. Kouba, M. Menaa, M. Hasni, M. Boudour, Load Frequency Control in multi-area power system based on Fuzzy Logic-PID Controller, in *2015 IEEE International Conference on Smart Energy Grid Engineering (SEGE)*, Oshawa, ON, Canada, 2015.
5. J. Pahasa, I. Ngamroo, Simultaneous control of frequency fluctuation and battery SOC in a smart grid using LFC and EV controllers based on optimal MIMO-MPC. J. Electr. Eng. Technol. **12**(2), 601–611 (2017)
6. H. Gozde, M.C. Taplamacioglu, Automatic generation control application with craziness based particle swarm optimization in a thermal power system. Int. J. Elec. Power. **33**(1), 8–16 (2011)
7. N. Ramesh Babu, A. Pachiyappan, Wind energy conversion systems-a technical review. J. Eng. Sci. Technol. **8**(4), 493–507 (2013)
8. H. Bevrani, *Robust power system frequency control* (Springer, New York, 2014)
9. Electrical Power Systems Technology, *Technical Guide: Basic Calculations for LV and HV Electrical Networks* (2010), http://eps-technology.blogspot.com/2010/12/technical-guide-basic-calculations-for.html.
10. E. Rakhshani, K. Rouzbehi, S. Sadeh, A new combined model for simulation of mutual effects between LFC and AVR loops, in *2009 Asia-Pacific Power and Energy Engineering Conference*, Wuhan, China (2009).
11. C.-T. Pan, C.-M. Liaw, An adaptive controller for power system load-frequency control. IEEE Trans. Power Syst. **4**(1), 122–128 (1989)
12. L. Jiang, W. Yao, Q.H. Wu, J.Y. Wen, S.J. Cheng, Delay-dependent stability for load frequency control with constant and time-varying delays. IEEE Trans. Power Syst. **27**(2), 932–941 (2012)
13. S. Doolla, T. Bhatti, Load frequency control of an isolated small-hydro power plant with reduced dump load. IEEE Trans. Power Syst. **21**(4), 1912–1919 (2006)
14. J.I. Pérez-Díaz, J.I. Sarasúa, Failures during load-frequency control maneuvers in an upgraded hydropower plant: causes, identification of causes and solution proposals. Energies **8**(10), 10584–10604 (2015)

G. Magdy et al., *Renewable Power Systems Dynamic Security*, Power Systems,
https://doi.org/10.1007/978-3-030-33455-0

15. K.S. Parmara, S. Majhi, D.P. Kothari, Load frequency control of a realistic power system with multi-source power generation. Int. J. Elec. Power. **42**(1), 426–433 (2012)
16. B. Singh, G.K. Kasal, Voltage and frequency controller for a three-phase four-wire autonomous wind energy conversion system. IEEE Trans. Energy Convers. **23**(2), 509–518 (2008)
17. D. Chen, S. Kumar, M. York, L. Wang, Smart automatic generation control, in *Proceedings of the 2012 IEEE Power and Energy Society General Meeting*, San Diego, CA, USA (2012).
18. A. Bose, I. Atiyyah, Regulation error in load frequency control. IEEE Trans. Power App. Syst. **PAS-99**(2), 650–657 (1980)
19. İ. Kocaarslan, E. Çam, Fuzzy logic controller in interconnected electrical power systems for load-frequency control. Int. J. Elec. Power. **27**(8), 542–549 (2005)
20. R. Verma, S. Pal, Sathans S, Intelligent automatic generation control of two-area hydrothermal power system using ANN and Fuzzy logic, in *2013 International Conference on Communication Systems and Network Technologies*, Gwalior, India (2013).
21. N. Ibraheem, H. Nizamuddin, T.S. Bhattib, AGC of two area power system interconnected by AC/DC links with diverse sources in each area. Int. J. Elec. Power. **55**, 297–304 (2014)
22. V.K. Kamboj, K. Arora, P. Khurana, Automatic generation control for interconnected hydrothermal system with the help of conventional controllers. Int. J. Electr. Comput. Eng. **2**(4), 547–552 (2012)
23. C.-F. Lu, C.-C. Liu, C.-J. Wu, Effect of battery energy storage system on load frequency control considering governor deadband and generation rate constraint. IEEE Trans. Energy Convers. **10**(3), 555–561 (1995)
24. I. Ngamroo, Y. Mitani, K. Tsuji, Application of SMES coordinated with solid-state phase shifter to load frequency control. IEEE Trans. Appl. Superconduct. **9**(2), 322–325 (1999)
25. K.R. Sudha, R. Santhi, Load frequency control of an interconnected reheat thermal system using type-2 fuzzy system including SMES units. Int. J. Elec. Power. **43**(1), 1383–1392 (2012)
26. H. Alhelou, M. Golshan, J. Askari-Marnani, Robust sensor fault detection and isolation scheme for interconnected smart power systems in presence of RER and EVs using unknown input observer. Int. J. Elec. Power. **99**, 682–694 (2018)
27. E. Cam, G. Gorel, H. Mamur, Use of the genetic algorithm-based fuzzy logic controller for load-frequency control in a two area interconnected power system. Appl. Sci. **7**(3), 1–22 (2017)
28. E. Davison, N. Tripathi, The optimal decentralized control of a large power system: load and frequency control. IEEE Trans. Automat. Contr. **23**(2), 312–325 (1978)
29. M. Rahmani, N. Sadati, Hierarchical optimal robust load-frequency control for power systems. IET Gener. Transm. Distrib. **6**(4), 303–312 (2012)
30. M. Shiroei, M.R. Toulabi, A.M. Ranjbar, Robust multivariable predictive based load frequency control considering generation rate constraint. Int. J. Elec. Power. **46**, 405–413 (2016)
31. H. Bevrani, F. Habibi, P. Babahajyani, M. Watanabe, Y. Mitani, Intelligent frequency control in an AC microgrid: online PSO-based fuzzy tuning approach. IEEE Trans. Smart. Grid. **3**(4), 1935–1944 (2012)
32. H. Bevrani, Y. Mitani, K. Tsuji, Robust decentralised load-frequency control using an iterative linear matrix inequalities algorithm. IET Proc. Generat. Transm. Distrib. **151**(3), 347–354 (2004)
33. L. Dong, Y. Zhang, Z. Gao, A robust decentralized load frequency controller for interconnected power systems. ISA Trans. **51**(3), 410–419 (2012)
34. T.H. Mohamed, J. Morel, H. Bevrani, T. Hiyamac, Model predictive based load frequency control_design concerning wind turbines. Int. J. Elec. Power. **43**(1), 859–867 (2012)
35. W. Tan, H. Zhou, Robust analysis of decentralized load frequency control for multi-area power systems. Int. J. Elec. Power. **43**(1), 996–1005 (2012)
36. O. Malik, A. Kumar, G. Hope, A load frequency control algorithm based on a generalized approach. IEEE Trans. Power Syst. **3**(2), 375–382 (1988)

37. S. Prakash, A.K. Bhardwaj, S.K. Sinha, Neuro fuzzy hybrid intelligent approach for four -area load frequency control of interconnected power system, in *2012 2nd International Conference on Power, Control and Embedded Systems*, Allahabad, India (2012).
38. K. Vrdoljak, N. Perić, I. Petrović, Sliding mode based load-frequency control in power systems. Electr. Pow. Syst. Res. **80**(5), 514–527 (2010)
39. O. Anaya-Lara, F. Hughes, N. Jenkins, G. Strbac, Contribution of DFIG-based wind farms to power system short-term frequency regulation. IET Proc. Generat. Transm. Distrib. **153**(2), 164–170 (2006)
40. G. Ramtharan, J. Ekanayake, N. Jenkins, Frequency support from doubly fed induction generator wind turbines. IET Renew. Power Gener. **1**(1), 3–9 (2007)
41. R. Doherty, A. Mullane, G. Nolan, D.J. Burke, A. Bryson, M. O'Malley, An assessment of the impact of wind generation on system frequency control. IEEE Trans. Power Syst. **25**(1), 452–460 (2010)
42. P.K. Ray, S.R. Mohanty, N. Kishor, A. Mohanty, Small-signal analysis of hybrid distributed generation system with HVDC-link and energy storage elements, in *2009 Second International Conference on Emerging Trends in Engineering & Technology*, Nagpur, India (2009).
43. H. Bevrani, B. Francois, T. Ise, *Microgrid dynamics and control* (John Wiley& Sons, Hoboken, NJ, 2017)
44. T. Kerdphol, F.S. Rahman, Y. Mitani, M. Watanabe, S. Küfeoğlu, Robust virtual inertia control of an islanded microgrid considering high penetration of renewable energy. IEEE Access **6**, 625–636 (2017)
45. T. Kerdphol, F.S. Rahman, Y. Mitani, K. Hongesombut, S. Küfeoğlu, Virtual inertia control-based model predictive control for microgrid frequency stabilization considering high renewable energy. Sustainability **9**(5), 1–21 (2017)
46. R. Yan, T.K. Saha, Frequency response estimation method for high wind penetration considering wind turbine frequency support functions'. IET Renew. Power Gener. **9**(7), 775–782 (2015)
47. A. Keyhani, A. Chatterjee, Automatic generation control structure for smart power grids. IEEE Trans. Smart. Grid. **3**(3), 1310–1316 (2012)
48. S. Vachirasricirikul, I. Ngamroo, Robust LFC in a smart grid with wind power penetration by coordinated V2G control and frequency controller. IEEE Trans. Smart. Grid. **5**(1), 371–380 (2014)
49. J.M. Guerrero, J.C. Vasquez, J. Matas, L.G. de Vicuna, M. Castilla, Hierarchical control of droop-controlled AC and DC microgrids—a general approach toward standardization. IEEE Trans. Ind. Electron. **58**(1), 158–172 (2011)
50. Siemens USA, Microgrids Solutions, https://w3.usa.siemens.com/smartgrid/us/en/microgrid/pages/microgrids.aspx
51. Smart Energy International, Ontario opens 2018 round of Smart Grid Fund (2018), https://www.smart-energy.com/industry-sectors/business-finance-regulation/grid-fund-ontario-energy-ministry/
52. R. Karasu, I.A. Chidambaram, PI2 controller based coordinated control with redox flow battery and unified power flow controller for improved restoration indices in a deregulated power system. Ain Shams Eng. J. **7**(4), 1011–1027 (2016)
53. L.C. Saikia, A. Bharali, O. Dixit, T. Malakar, B. Sharma, S. Kouli, Automatic generation control of multi-area hydro system using classical controllers, in *2012 1st International Conference on Power and Energy in NERIST (ICPEN)*, Nirjuli, India (2012).
54. P. Topno, S. Chanana, Load frequency control of a two-area multi-source power system using a tilt integral derivative controller. J. Vib. Control. **24**(1), 110–125 (2016)
55. M. Aoki, Control of large-scale dynamic systems by aggregation. IEEE Trans. Automat. Contr. **13**(3), 246–253 (1968)
56. Y.-H. Moon, H.-S. Ryu, B. Kim, K.-B. Song, Optimal tracking approach to load frequency control in power systems, in *2000 IEEE Power Engineering Society Winter Meeting. Conference Proceedings (Cat. No.00CH37077)*, Singapore, Singapore (2000).

57. M. Farahani, S. Ganjefar, M. Alizadeh, PID controller adjustment using chaotic optimisation algorithm for multi-area load frequency control. IET Control Theory Appl. **6**(12), 1984–1992 (2012)
58. M.A. Tammam, M.A.S. Aboelela, M.A. Moustafa, A.E.A. Seif, Fuzzy like PID controller tuning by multi-objective genetic algorithm for load frequency control in nonlinear electric power systems. Int. J. Adv. Eng. Technol. **5**(1), 572–583 (2012)
59. C.E. Fosha, O.I. Elgerd, The megawatt-frequency control problem: a new approach via optimal control theory. IEEE Trans. Power App. Syst. **PAS-89**(4), 563–577 (1970)
60. M. Calovic, Linear regulator design for a load and frequency control. IEEE Trans. Power App. Syst. **PAS-91**(6), 2271–2285 (1972)
61. V. Moorthi, Damping effects of excitation control in load-frequency control system. Proc. Instit. Electr. Eng. **21**(11), 1409–1416 (1974)
62. R. Shoults, Multi-area adaptive LFC developed for a comprehensive AGC simulator. IEEE Trans. Power Syst. **8**(2), 541–547 (1993)
63. K.A. Lee, H. Yee, C.Y. Teo, Self-tuning algorithm for automatic generation control in an interconnected power system. Electr. Power Syst. Res. **20**(3), 157–165 (1991)
64. S. Jovanovic, Self-tuning steam turbine controller in a multi-machine power system. IEEE Trans. Energy Convers. **14**(4), 1578–1581 (1999)
65. Y. Wang, R. Zhou, C. Wen, Robust load-frequency controller design for power systems. IET Proc. Generat. Transm. Distrib. **140**(1), 11–16 (1993)
66. V.P. Singhb, S.R. Mohanty, N. Kishor, P.K. Rayc, Robust H-infinity load frequency control in hybrid distributed generation system. Int. J. Elec. Power. **46**, 294–305 (2013)
67. H. Bevrani, M.R. Feizi, S. Ataee, Robust frequency control in an Islanded Microgrid:$H\infty$ and µ-synthesis approaches. IEEE Trans. Smart. Grid. **7**(2), 706–717 (2016)
68. H.A. Yousef, K. AL-Kharusi, M.H. Albadi, N. Hosseinzadeh, Load frequency control of a multi-area power system: an adaptive fuzzy logic approach. IEEE Trans. Power Syst. **29**(4), 1822–1830 (2014)
69. R. Ali, T.H. Mohamed, Y.S. Qudaih, Y. Mitanib, A new load frequency control approach in an isolated small power systems using coefficient diagram method. Int. J. Elec. Power. **56**, 110–116 (2014)
70. P. Garasi, Y. Qudaih, R. Ali, M. Watanabe, Y. Mitani, Coefficient diagram method-based load frequency control for a modern power system. J. Electr. Sci. Technol. **12**(3), 270–276 (2014)
71. T.H. Mohamed, H. Bevrani, A.A. Hassan, T. Hiyama, Decentralized model predictive based load frequency control in an interconnected power system. Energ. Conver. Manage. **52**(2), 1208–1214 (2011)
72. G. Shabib, E. Abd-Elhameed, G. Magdy, Plant input mapping digital redesign of a PID controller for a power system damping, in *3rd International Conference on Energy Systems and Technologies*, Cairo, Egypt (2015).
73. P. Bhatt, R. Roy, S.P. Ghoshal, Dynamic participation of doubly fed induction generator in automatic generation control. Renew. Energy **36**(4), 1203–1213 (2011)
74. H.M. Hasanien, A.A. El-Fergany, Symbiotic organisms search algorithm for automatic generation control of interconnected power systems including wind farms. IET Gener. Transm. Distrib. **11**(7), 1692–1700 (2017)
75. P. Xie, Y. Li, L. Zhu, D. Shi, X. Duan, Supplementary automatic generation control using controllable energy storage in electric vehicle battery swapping stations. IET Gener. Transm. Distrib. **10**(4), 1107–1116 (2016)
76. G. El-Saady, Y. Mobarak, M.A.A. Wahab, M.M. Hamada, M. Abdel-Moamen, A.M. El-Noby, A new robust technique LFC of multi-area power system using SMES. Int. J. Control Auto. Syst. **1**(2), 56–60 (2013)
77. P. Bhatt, R. Roy, S.P. Ghoshal, Dynamic contribution of DFIG along with SMES for load frequency control of interconnected restructured power system, in *2011 10th International Conference on Environment and Electrical Engineering*, Rome, Italy (2011).
78. A.-A.A. Mohamed, Y.S. Mohamed, A.A.M. El-Gaafary, A.M. Hemeida, Optimal power flow using moth swarm algorithm. Electr. Pow. Syst. Res. **142**, 190–206 (2017)

79. Egyptian Electricity Holding Company, Annual Report, Cairo, Egypt (2016), http://www.eehc.gov.eg/eehcportal/YearlyReports.aspx.
80. H. Mostafa, Load frequency control of the Egyptian power system using bacterial foraging algorithm. Port Said Eng. Res. J. **16**(2), 1–10 (2012)
81. G. Magdy, E.A. Mohamed, G. Shabib, A.A. Elbaset, Y. Mitani, SMES based a new PID controller for frequency stability of a real hybrid power system considering high wind power penetration. IET Renew. Power Gener. **12**(11), 1304–1313 (2018)
82. Mubashar yaqoob Zargar, Mairaj-ud-Din Mufti, Shameem Ahmed Lone, Voltage and frequency control of a hybrid wind-diesel system using SVC and predictively controlled SMES, in *2017 6th International Conference on Computer Applications In Electrical Engineering-Recent Advances (CERA)*, Roorkee, India (2017).
83. H.M. Hasanien, Whale optimisation algorithm for automatic generation control of interconnected modern power systems including renewable energy sources. IET Gener. Transm. Distrib. **12**(3), 607–614 (2018)
84. B.J. Brearley, R. RajaPrabu, A review on issues and approaches for microgrid protection. Renew. Sustain. Energy Rev. **67**, 988–997 (2017)
85. H. Bevrani, T. Ise, Y. Miurab, Virtual synchronous generators: a survey and new perspectives. Int. J. Elec. Power. **54**, 244–254 (2014)
86. T. Keil, J. Jager, Advanced coordination method for overcurrent protection relays using nonstandard tripping characteristics. IEEE Trans. Power Delivery **23**(1), 52–57 (2008)
87. S. Kunsman, S. Meier, R. Hedding, Protection and control system impacts from The Digital World, in *2016 69th Annual Conference for Protective Relay Engineers (CPRE)*, College Station, TX, USA (2016).
88. G. Shabib, Digital design of a power system stabilizer for a power system based on plant input mapping. Int. J. Elec. Power. **49**, 40–46 (2013)
89. G. Parise, L. Martirano, M. Kermani, M. Kermani, Designing a power control strategy in a microgrid using PID/fuzzy controller based on battery energy storage, in *2017 IEEE International Conference on Environment and Electrical Engineering and 2017 IEEE Industrial and Commercial Power Systems Europe (EEEIC/I&CPS Europe)*, (2017), pp. 1–5.
90. G. Shabib, N. Hori, Discrete time models of a continuous power system stabilizer, in *SICE Annual Conference 2007*, Takamatsu, Japan (2007).
91. N. Rafee, T. Chen, O. Malik, A technique for optimal digital redesign of analog controllers. IEEE Trans. Control Syst. Technol. **5**(1), 89–99 (1997)
92. E. Sortomme, G.J. Mapes, B.A. Foster, S.S. Venkata, Fault analysis and protection of a microgrid, in *2008 40th North American Power Symposium, Calgary, AB, Canada*, (2008)
93. S. Sheng, K.K. Li, W.L. Chan, X. Zeng, D. Shi, X. Duan, Adaptive agent-based wide-area current differential protection system. IEEE Trans. Ind. Appl. **46**(5), 2111–2117 (2010)
94. J.A. Laghari, H. Mokhlis, A.H.A. Bakar, H. Mohamad, Application of computational intelligence techniques for load shedding in power systems: a review. Energ. Conver. Manage. **75**, 130–140 (2013)
95. N. Tephiruk, K. Hongesombut, Y. Urathamakul, S. Poonvasin, S. Tangsatit, Modeling of rate of change of under frequency relay for microgrid protection, in *2017 International Electrical Engineering Congress (iEECON)*, Pattaya, Thailand (2017).
96. J. Vieira, W. Freitas, W. Xu, A. Morelato, Efficient coordination of ROCOF and frequency relays for distributed generation protection by using the application region. IEEE Trans. Power Del. **21**(4), 1878–1884 (2006)
97. E.A. Mohamed, G. Magdy, G. Shabib, A.A. Elbaset, Y. Mitani, Digital coordination strategy of protection and frequency stability for an islanded microgrid. IET Gener. Transm. Distrib. **12** (15), 3637–3646 (2018)
98. A.A.M. Hassan, T.A. Kandeel, Effectiveness of frequency relays on networks with multiple distributed generation. J. Electr. Syst. Inf. Technol **2**(1), 75–85 (2015)
99. S.F. Zarei, M. Parniani, A comprehensive digital protection scheme for low-voltage microgrids with inverter-based and conventional distributed generations. IEEE Trans. Power Del. **32**(1), 441–452 (2017)

100. M.M. Aman, G.B. Jasmon, H.B. Mokhlis, Q.A. Khan, A.H.B.A. Bakar, M. Karimi, Modeling and simulation of digital frequency relay for generator protection, in *2012 IEEE International Conference on Power and Energy (PECon)*, Kota Kinabalu, Malaysia (2012).
101. European Network of Transmission System Operators for, Electricity (entsoe), Supporting document for the network code on load-frequency control and reserves, in Chapter 4 (2013), pp. 17–36.
102. G. Magdy, G. Shabib, A.A. Elbaset, Y. Mitani, A novel coordination scheme of virtual inertia control and digital protection for microgrid dynamic security considering high renewable energy penetration. IET Renew. Power Gener. **13**(3), 462–474 (2019)
103. T. Kerdphol, F.S. Rahman, Y. Mitani, Virtual inertia control application to enhance frequency stability of interconnected power systems with high renewable energy penetration. Energies **11** (4), 1–16 (2018)
104. K. Mentesidi, R. Garde, M. Aguado, E. Rikos, Implementation of a fuzzy logic controller for virtual inertia emulation, in *2015 International Symposium on Smart Electric Distribution Systems and Technologies (EDST), Vienna, Austria*, (2015)
105. Y. Hu, W. Wei, Y. Peng, J. Lei, Fuzzy virtual inertia control for virtual synchronous generator, in *2016 35th Chinese Control Conference (CCC), Chengdu, China*, (2016)
106. P.F. Frack, P.E. Mercado, M.G. Molina, Extending the VISMA concept to improve the frequency stability in Microgrids, in *2015 18th International Conference on Intelligent System Application to Power Systems (ISAP)*, Porto, Portugal (2015).
107. F. Gonzalez-Longatt, E. Chikuni, E. Rashayi, Effects of the Synthetic Inertia from wind power on the total system inertia after a frequency disturbance, in *2013 IEEE International Conference on Industrial Technology (ICIT)*, Cape Town, South Africa (2013).
108. W. Freitas, W. Xu, C. Affonso, Z. Huang, Comparative analysis between ROCOF and vector surge relays for distributed generation applications. IEEE Trans. Power Del. **20**(2), 1315–1324 (2005)
109. M.H. Khooban, T. Niknam, F. Blaabjerg, T. Dragičević, A new load frequency control strategy for micro-grids with considering electrical vehicles. Electr. Pow. Syst. Res. **143**, 585–598 (2017)
110. J. Kennedy, R. Eberhart, Particle swarm optimization, in *Proceedings of ICNN'95—International Conference on Neural Networks*, Perth, WA, Australia (1995).
111. T. Kerdphol, Y. Qudaih, Y. Mitani, Optimum battery energy storage system using PSO considering dynamic demand for microgrids. Int. J. Elec. Power. **83**, 58–66 (2016)
112. K.H. Ang, G. Chong, Y. Li, PID control system analysis, design, and technology. IEEE Trans. Control Syst. Technol. **13**(4), 559–576 (2005)
113. Z.-L. Gaing, A particle swarm optimization approach for optimum design of PID controller in AVR system. IEEE Trans. Energy Convers. **19**(2), 384–391 (2004)
114. Australian Energy Market Commission (AEMC), Reliability and Panel, Application of frequency operation standards during periods of supply scarcity, *Final Determination*, Sydney, Australia (2009).
115. H. Wu, X. Ruan, D. Yang, X. Chen, W. Zhao, Z. Lv, Q.-C. Zhong, Small-signal modeling and parameters design for virtual synchronous generators. IEEE Trans. Ind. Electron. **63**(7), 4292–4303 (2016)
116. I. Cvetkovic, D. Boroyevich, R. Burgos, C. Li, M. Jaksic, P. Mattavelli, Modeling of a virtual synchronous machine-based grid-interface converter for renewable energy systems integration, in *2014 IEEE 15th Workshop on Control and Modeling for Power Electronics (COMPEL)*, Santander, Spain (2014).
117. A. Fathi, Q. Shafiee, H. Bevrani, Robust frequency control of microgrids using an extended virtual synchronous generator. IEEE Trans. Power Syst. **33**(6), 6289–6297 (2018)
118. D. Li, Q. Zhu, S. Lin, X.Y. Bian, A self-adaptive inertia and damping combination control of VSG to support frequency stability. IEEE Trans. Energy Convers. **32**(1), 397–398 (2017)
119. J. Liu, Y. Miura, T. Ise, Comparison of dynamic characteristics between virtual synchronous generator and droop control in inverter-based distributed generators. IEEE Trans. Power Electron. **31**(5), 3600–3611 (2016)

120. Y. Hirase, K. Abe, K. Sugimoto, K. Sakimoto, H. Bevrani, T. Ised, A novel control approach for virtual synchronous generators to suppress frequency and voltage fluctuations in microgrids. Appl. Energy **210**, 699–710 (2018)
121. J. Liu, Y. Miura, H. Bevrani, T. Ise, Enhanced virtual synchronous generator control for parallel inverters in microgrids. IEEE Trans. Smart. Grid. **8**(5), 2268–2277 (2017)
122. H.C. Kılıçkıran, İ. Şengör, H. Akdemir, B. Kekezoglu, H. Akdemir, O. Erdinc, N.G. Paterakis, Power system protection with digital overcurrent relays: a review of non-standard characteristics. Electr. Power Syst. Res. **164**, 89–102 (2018)
123. M. Singh, V. Telukunta, S.G. Srivanib, Enhanced real time coordination of distance and user defined over current relays. Int. J. Elec. Power. **98**, 430–441 (2018)
124. M. Dreidy, H. Mokhlis, S. Mekhilef, Inertia response and frequency control techniques for renewable energy sources: a review. Renew. Sustain. Energy Rev. **69**, 144–155 (2017)
125. G. Magdy, G. Shabib, A.A. Elbaset, Y. Mitani, Renewable power systems dynamic security using a new coordination of frequency control strategy based on virtual synchronous generator and digital frequency protection. Int. J. Elec. Power. **109**, 351–368 (2019)
126. K. Valarmathi, D. Devaraj, T. Radhakrishnan, Particle Swarm Optimization based PI controller tuning for Fermentation Process, in *International Conference on Computational Intelligence for Modelling, Control and Automation and International Conference on Intelligent Agents, Web Technologies and Internet Commerce (CIMCA-IAWTIC'06)*, Vienna, Austria (2005).
127. Y. Qudaih, I. Moukhtar, T. H. Mohamed, A. A.Elbaset, A. Z. ElDein and Y. Mitani, "Parallel PI/CDM frequency controller to support V2G plan for microgrid," Energy Procedia, vol. 100, pp. 342-351, 2016.
128. Egyptian Electricity Holding Company, Annual Report 2017, Cairo, Egypt (2017), http://www.moee.gov.eg/english_new/report.aspx.
129. P. Kundur, *Power system stability and control*, 5th edn. (McGraw-Hill Education, New Delhi, 2008)
130. S. Saxen, Y.V. Hote, Decentralized PID load frequency control for perturbed multi-area power systems. Int. J. Elec. Power. **81**, 405–415 (2016)
131. Y. Sun, Y. Wang, Z. Wei, G. Sun, X. Wu, Robust H∞ load frequency control of multi-area power system with time delay: a sliding mode control approach. IEEE/CAA J. Automat. Sin. **5** (2), 610–617 (2018)
132. L. Zhang, S. Wang, H.R. Karimi, A. Jasra, Robust finite-time control of switched linear systems and application to a class of servomechanism systems. IEEE/ASME Trans. Mechatron. **20**(5), 2476–2485 (2015)
133. H. Bevrani, T. Hiyamab, Robust decentralised PI based LFC design for time delay power systems. Energ. Conver. Manage. **49**(2), 193–204 (2008)
134. Q. Zhu, L. Jiang, W. Yao, C. Zhang, C. Luo, Robust load frequency control with dynamic demand response for deregulated power systems considering communication delays. Electr. Power Comp. Syst. **45**(1), 75–87 (2017)
135. P.P. Khargonekar, N. Sivashankar, H_2 optimal control for sampled-data systems. Syst. Contr. Lett. **17**(6), 425–436 (1991)
136. J. Keller, B. Anderson, A new approach to the discretization of continuous-time controllers. IEEE Trans. Automat. Contr. **37**(2), 214–223 (1992)
137. C. Rabbath, N. Hori, Reduced-order PIM methods for digital redesign. IEE Proc. Contr. Theory Appl. **150**(4), 335–346 (2003)
138. K. Ogata, *Discrete-time control system* (Prentice Hall, Upper Saddle River, NJ, 1995)
139. Global Wind Energy Council (GWEC), Global wind report annual market update (2018), http://www.gwec.net.
140. G. Magdy, G. Shabib, A.A. Elbaset, T. Kerdphol, Y. Qudaih, H. Bevrani, Y. Mitani, Tustin's technique based digital decentralized load frequency control in a realistic multi power system considering wind farms and communications delays. Ain Shams Eng. J. **10**(2), 327–341 (2019)
141. The Wind Power, Wind energy database, https://www.thewindpower.net/turbine_en_42_gamesa_g52-850.php.

Index

G. Magdy et al., *Renewable Power Systems Dynamic Security*, Power Systems,
https://doi.org/10.1007/978-3-030-33455-0

I

L

M

N

O

P

Printed by Printforce, the Netherlands